W0043796

Uwe Harms (Ed.)

Supercomputer and Chemistry 2

debis Workshop 1990

Ottobrunn, November 19–20, 1990

With 80 Figures and 21 Tables

Springer-Verlag Berlin Heidelberg GmbH

Uwe Harms

debis Systemhaus GmbH
LRZ Ottobrunn
Leibnizstraße 7, W-8012 Ottobrunn
Federal Republic of Germany

ISBN 978-3-540-54411-1   ISBN 978-3-642-50175-3 (eBook)
DOI 10.1007/978-3-642-50175-3

© Springer-Verlag Berlin Heidelberg 1991
Originally published by Springer-Verlag Berlin Heidelberg New York in 1991
Softcover reprint of the hardcover 1st edition 1991

Typesetting: Camera ready by author
51/3140-543210 – Printed on acid-free paper

Further developments for the IGLO (Individual gauge for localized orbitals) program, which was discussed in the 1989 workshop, are presented. The NMR chemical shifts are calculated by using TURBOMOLE: A new and very interesting type of parallelism is proposed: the dataflow parallelism. The results are shown in an optimized run with pipelined input/output.

Finally, modelling studies of the three dimensional structures of the saruplase-domains are discussed. They help in computer-aided protein design, as crystallography X-ray has failed. The programs used are WHATIF and GROMOS; a minisupercomputer delivering the computing power.

This year the very short preparation time presented several problems but they were solved with the support of Dr. R. Iffert and N. McCann. This successful and fruitful cooperation will be needed again for the workshop 1991 that is planned for November.

April 1990   U. Harms

Preface

Supercomputer and Chemistry, a debis workshop continues the serie of seminars organized by IABG until 1989. In summer of 1990, debis Systemhaus GmbH (Daimler-Benz Interservices) acquired the IABG computer center including the Supercomputer Siemens VP200 and the personnel. However, we managed to organize the 1990 workshop at such short notice.

This year we again combined the aspects of supercomputing, computational sciences and computer-aided chemistry.

First, the current situation in the world of supercomputing is highlighted. Some new information concerning the supercomputers and their distribution in Japan has been collected. In the course of 1990, several supercomputer centers in Germany replaced their machines. A newly compended list is contained below.

An important influence on the usage of supercomputers is networking. The state of the art in local and wide area networks is described. Two manufacturers of high speed networks give some examples of the structure of networks at present and in the future.

The main part of the workshop was dedicated to computational chemistry. A direct SCF was parallelized and implemented on a set of RISC-workstations, an interesting and cost-effective solution. First results were presented.

The proven importance of computational methods complementing experimental approaches for studying short-lived intermediates - carbocations and alkyl radicals - is demonstrated.

For long-time dynamics of proteins a new method is proposed; it is based on the Monte Carlo method with a window algorithm. Furthermore, a general overview on quantum chemical calculations of small molecules as a predictive and analytical tool for chemical research is presented. The importance is demonstrated by several research projects.

Different types of parallel operations on different levels are discussed. They result in new compiler techniques for auto-parallelization. The performance and the limitations are demonstrated by several application programs for computational chemistry.

Table of Contents

Supercomputing – What is new

U. Harms

debis Systemhaus, Leibnizstraße 7, W-8012 Ottobrunn, FRG

Abstract: The current situation in the supercomputer market is discussed. New manufacturers try to market massively parallel systems. In 1990, several new subsidiaries were founded. On the other hand, the German supercomputer project SUPRENUM was cancelled. As there are no new interesting systems compared to the workshop in 1989, the situation in Japan is discussed in some detail, although the information sources differ in figures, systems and companies. Finally, a list of supercomputer installations in Germany is provided.

1. SUPERCOMPUTING WORLDWIDE

In 1990, the situation worldwide has not changed significantly. The growth in the number of systems has slowed down. Compared to a growth of nearly 35% from 1988 to 1989 now we saw only 2% increase in systems. That means, that in 1990 we have about 415 supercomputers compared to 404 in 1989 and 300 in 1988.

Broken down, there are about 95 systems in Europe of which 25 are in France, 26 in Germany, 22 in Great Britain. In Japan we counted 138 Supercomputers and 171 in the USA and Canada.

Comparing the vendors, Cray has a worldwide market share of about 60%, the next is Fujitsu (Siemens AG/Siemens-Nixdorf, Amdahl) with about 20%, followed by Hitachi and NEC with 7%. The CDC/ETA figures are going down, as these systems are no longer produced.

2. PARALLEL PROCESSING

An increasing number of companies have gone massively parallel. This year in Germany Ncube, Thinking Machines and MassPar have founded subsidiaries. BBN is selling its machines from Italy.

A very well known transputer based on parallel systems is built and sold by Parsytec in Germany. They can connect a nearly unlimited number of transputers to a big machine.

Last year, still in the development phase, IP Systems has now shown its first parallel system, based on the Intel i860 processor. It has a tree architecture, so the communication can be handled very easily. They call their idea "wave concept", the wave is going down to the leaves and is then reflected with the results to the top.

One year ago, SUPRENUM installed its first 256 processor system at GMD. The experiences were not as good as expected. Hard- and software problems, e.g., compilers and the processor speed gave reasons to cancel the project at end of 1990. Five systems have been delivered, to the University of Liverpool, to KFA Jülich, University of Erlangen, Suprenum itself and the GMD.

U. Harms (Ed.)
Supercomputer and Chemistry 2
© Springer-Verlag Berlin Heidelberg 1991

The company has now been split into SUPRENUM, which will maintain the existing systems and PALLAS which will cooperate with Meiko and will develop or port software on parallel machines.

Further on, there is Intel with its iPSC/860 Hypercube.

Parallelism is growing rapidly but the application software is not keeping pace. The autoparallelising compilers need more know how and intelligence and are still in the nursing phase. Some experiences with the Alliant compiler can be found in the presentation of P. Weiner.

One limiting factor in massively parallel system is Amdahls' law, which maintains that the sequential part of a program dominates the parallel. Even with a degree of parallelization of 99%, the speed-up is limited by the factor of 100. Another problem is the communication between processors. So there is a lot to be done before massively parallel systems can be used effectively.

3. SUPERCOMPUTING IN JAPAN

In the meantime, the Japanese Supercomputer vendors are attacking Cray Research with their shared-memory parallel systems (Fujitsu, NEC). The Japanese are also planning - like Cray - massively parallel systems.

According to S. Jarp, the installation base in Japan is as follows:

Table 1: Japans Supercomputers by Vendors

	customers	including vendors
Fujitsu	63	73
Hitachi	18	29
Cray	26	26
NEC	18	21
CDC/ETA	2	2
Total	129	153

Three Cray Y/MP2E are on order and will be installed in the first quarter of 1991.

S. Jarp lists all the installations and their main applications, so that they can be ordered in the following way.

Table 2: Japanese Supercomputers by Usage and Vendors

	Cray	Fujitsu	Hitachi	NEC
University	2	16	3	7
Gov./Res. Labs	2	12	·4	3
Automobil	12	3	2	1
Service bureaus	5	3	4	2
Mechanical Engineering	1	9		4
Chemistry	2	5	2	
Financial		1	1	1
Optical		1		
Consulting		1		
Electronics		1		
Conglomerate	5	8	2	

ETA-Systems has installed its supercomputers only in universities, Meiji-University and Tokyo Institut of Technology.

In this context it is quite interesting to list the systems in chemical applications, that can be recognized.

Table 3: Chemical Applications

Asahi Chemical, Shizuoka, Cray Y-MP2E/116, March 1991
Sumitomo Chemical, Osaka, Cray X-MP/116se, September 1989
Chiyoda Info Service (?), Tokyo, Fujitsu VP 50, April 1986
Fuji Electro-Chemical (?), Tokyo, Fujitsu VP 50E, November 1988
Kodak Japan, Tokyo, Fujitsu VP 50E, November 1988
Mitsubishi Kasei, Kanagawa, Fujitsu VP 50, July 1986
Shionogi, Osaka, Fujitsu VP 30, May 1987
Toray, Tokyo, Fujitsu VP 30, August 1987
Bridgestone, Tokyo, Hitachi S 810/5, May 1987
Toyo Gum, Tokyo, Hitachi S810/5, October 1987
IMS (Molecular Science), Ibaraki, Hitachi S820/80, January 1988 (Gov. and Nat. Lab)

It should also be mentioned that there are about 110 IBM vector facilities, but it was not possible to get a detailed list.

4. SUPERCOMPUTING IN GERMANY

As mentioned above, there are no major changes in Germany. The first NEC machine was installed in Köln, as can be seen in table 6. There are about 54 vector facilities in Germany. This year there will be several new machines in Germany. The Universities of Darmstadt, Frankfurt, Kassel, Giessen are in the federal procurement procedure.

Rumors are heard from the Universities of Bremen and Dresden that they too will install supercomputers.

Table 4: Supercomputers in Research Institutes

	first install.	computer	date of replacements	computer
MPI Garching	1979	Cray 1	1986	Cray X-MP 2 (4 MWords)
KFA Jülich	1983	Cray X-MP 2	1989	Cray X-MP 4 (16 MWords)
DLR Oberpfaffenhofen	1983	Cray 1S	1987 1990	Cray X-MP 2 Cray Y-MP 2 (32 MWords)
Klimarechenzentrum Hamburg	1985	CDC Cyber 205	1988	Cray 2S (4 Proc) (128 MWords)
KfK Karlsruhe	1987	Siemens VP50	1990	Siemens VP400EX (64 MWords)
HLRZ Jülich	1987	Cray X-MP 4	1989	Cray Y-MP 8 (32 MWords)

Table 5: Supercomputers in Universities

	first install.	computer	date of replacements	computer
Bochum	1981	CDC Cyber 205		
Karlsruhe	1983	CDC Cyber 205	1988 1990 1991	Siemens VP400EX Siemens S400/10 Siemens S600/20 (128 MWords)
Stuttgart	1983	Cray 1M	1986	Cray 2D (4 proc) (256 MWords)
Berlin (ZIB)	1984	Cray 1M	1986	Cray X-MP 2 (4 MWords)
Kaiserslautern	1986	Siemens VP100 (16 MWords)		
Kiel	1987	Cray X-MP 1	1988	Cray X-MP 2 (16 MWords)
Munich (LRZ)	1988	Cray X-MP 2	1989	Cray Y-MP 4 (32 MWords)
Hanover (RRZN)	1989	Siemens VP200EX	1990 1991	Siemens S400/10 (64 MWords) Siemens S400/40 (128 MWords)
Aachen (RWTH)	1990	Siemens VP200EX	1991 1992	Siemens S400/10 (64 MWords) Siemens S600/20 (128 MWords)
Cologne	1990	NEC SX-3/11 (32 MWords)		

Table 6: Supercomputers in Industry

	first install.	computer	date of replacements	computer
Prakla-Seismos Hanover	1981	CDC Cyber 205	1989 1990	Cray X-MP 1 Cray Y-MP 2 (32 MWords)
debis Systemhaus Ottobrunn	1985	Siemens VP200 (32 MWords)		
EDS (Opel) Rüsselsheim	1985	Cray 1S	1989	Cray X-MP 1 (8 MWords)
Siemens Munich	1985	Siemens VP200 (16 MWords)	1991	Siemens S200/20 (64 MWords)
Gouvernment BMVG Bonn	1985	Cray X-MP 4 (8 MWords)		
debis Systemhaus Stuttgart	1987	Cray X-MP 2	1989	Cray Y-MP 2 (16 MWords)
Volkswagen Wolfsburg	1987	Cray X-MP 1	1989	Cray X-MP 2 (8 MWords)
Deutscher Wetter- dienst Offenbach	1987	CDC ETA-10	1990	Cray Y-MP 4 (32 MWords)
Siemens/Nixdorf Munich	1988	Siemens VP100EX (32 MWords)		
BMW Munich	1988	Cray X-MP 2 (8 MWords)		
Continental Gummi Hanover	1989	Cray X-MP 1 (8 MWords)		

References

Jarp, S.: A review of Japan und Japanese high-end Computers, Europ. Lab for Particel Physics, CERN Rep., CN/91/01, January 1991

Gentzsch, W.: private communication

Meuer, H.-W.: Einführung in das Seminar Supercomputer '90, Mannheim, 1990

Meuer, H.-W. (Ed.): Supercomputer '90, Mannheim, Informatik-Fachberichte 250, Springer-Verlag

Local Area Networks – A Survey

G. Michalk

Systemhaus DisCom, Distributed Computing GmbH, Roermonder Str. 615,
W-5100 Aachen-Richterich, FRG

Abstract: The OSI reference model of data communications will be described with respect
to local area networks (LANs). A short survey of the two most important, "conventional"
LAN architectures, Ethernet and Token-Ring, will be given together with some examples
of possible topologies and infrastructures. Then some of the principles and highlights of
the relatively new FDDI (Fiber Distributed Data Interface) high-speed LAN will be
reported. Finally existing and new types of network applications will be discussed.

THE OSI-MODEL

Today all the topics which are related to problems and solutions in the field of data
communications are discussed within the framework of the so called OSI reference model.
OSI stands for Open System Interconnection and is the result of the efforts of the
International Organisation for Standardization (ISO) which began in the mid '70s and
ended in 1984 in the ISO 7498 norm [1].
This norm, which was accepted by the European CCITT (X.200), describes the process of
data communication within a hierarchical, layered architecture model for data exchange. It
defines the components of this process and their relationship to one another.
The description is based on an seven layer protocol stack, where the top layer represents
the the user application program interface in the communication process. The bottom
layer handles the actual conversion of data which have been processed by all
intermediate layers into physical signals (voltages, light, etc.) which propagate through
some media (copper or fiber optic cable) towards the other partner (see fig. 1). One
"layer" represents a closely related group of functions. This functionality is offered as a
"service" only to the layers just below and above. A "protocol" is set of rules which govern
communications between the same layers of different partners. As indicated by fig. 1, the
top three layers are application-oriented in contrast to the four bottom layers which are
more transport-oriented.
From a strict OSI point of view, all specific details of a local area network (LAN) are
confined to the bottom three or even the last two layers of a network. The Physical Layer

U. Harms (Ed.)
Supercomputer and Chemistry 2
© Springer-Verlag Berlin Heidelberg 1991

(layer 1, sometimes also split into two sublayers (PHY (physical layer protocol) and PMD (physical media dependent)), transforms the digital information into physical signals on one side (voltages propagating on a copper cable or infrared radiation propagating in a fiber optic cable) and vice versa on the other side.

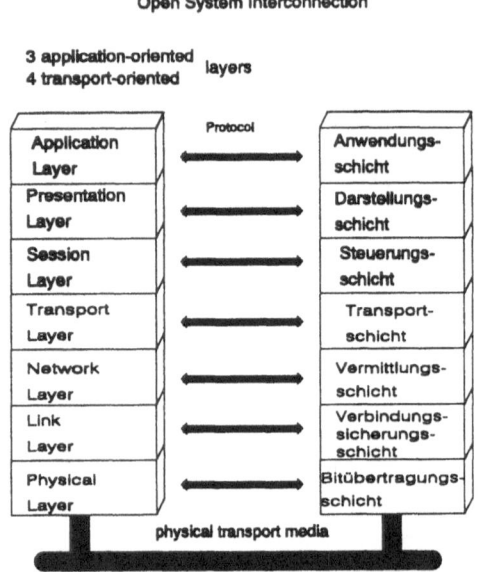

Fig. 1: OSI Reference Model

The Link Layer (layer 2) is responsible for a reliable transport of the data on a single network segment. It uses some kind of flow control and assures a correct and complete delivery of all data. In LANs, layer 2 is split into two sublayers: the Logical Link Control (LLC, IEEE 802.2) sublayer which offers media independent services to layer 3 and the Medium Access Control (MAC) sublayer, with its different implementations Ethernet (Ethernet V.: 2.0, IEEE 802.3), Token-Ring (IEEE 802.5), FDDI (ANSI X3T9.5) and others. The Network Layer (layer 3) implements a reliable connection between endsystems, eventually coupling the data flow between these systems over various layer 2 segments of different type and structure. Two workstations on two separated LANs with an intermediate X.25 connection between the LANs is an example of such a connection.

ETHERNET

Ethernet was developped originally by Xerox Corp. beginning in 1973, published in 1976, and standardized by Digital Equipment, Intel Corp. and Xerox Corp. (DIX-Ethernet) in two versions (1.0, 1980, and 2.0, 1982) [2]. The most widespread current protocol implementations (TCP/IP , DECNET Phase IV) are based on Ethernet V 2.0. This standard comprises a full implementation of the bottom two OSI layers. The international standards IEEE 802.3 and ISO 8802.3, which also cover the bottom 1 1/2 layers up to the MAC-layer of an Ethernet network are not compatible with the DIX V 2.0 standard (see [2] or [5] for details), but can coexist without interference, in most cases, on the same physical medium (cable or fiber optic) . Network protocols, that were later developed are more often based on the ISO standard (Novell IPX or IBMs NETBIOS).

Ethernet uses a bus topology with CSMA/CD (carrier sense multiple access with collision detection), independently of the underlying cabling system. This means, that all stations on an Ethernet have the same right to access the medium, and that the evolving conflicts from simultaneous acesses ("collisions") can be recognized and resolved using a special procedure that is implemented in all station adapters. A single Ethernet can accomodate up to 1024 stations. The bit data rate is 10 Mbps (million bits per second), with the actual data rate varying, depending on the specific processing conditions.

Nowadays Ethernet can be implemented on several quite different cabling structures. Historically, the oldest alternative is the "Yellow Cable", a thick coaxial cable, which allows for a maximal segment length of 500 m (10BASE5). A similar concept is used with the "Cheapernet" or "Thin wire" Ethernet, based on a 50 Ohm coaxial cable, with a maximum segment length of 185 m (10BASE2), but with easier and much cheaper access for a workstation to the network. Today, most Ethernet station adapters on the market support direct attachment to both types of Ethernet cables. In the last few years, an implementation of Ethernet based on star-wiring twisted pair (shielded like IBM's type 1 or unshielded voice-grade) cable also emerged, which is standardized as 10BASET variant. The typical radius of such a star-wired Ethernet is app. 100 m. For larger diameter or campus-wide networks, most manufacturers of networking equipment offer Ethernet implementations on fiber optic cables mostly in a star-wired topology. With these products, which are currently not standardized, network diameters of up to 4.5 kms can be spanned. All of these technologies can be mixed and integrated to migrate from a historical thick Ethernet-based structure to a structured cabling topology using twisted pair copper and fiber optic media (see fig. 2).

The earliest implementations of network protocols on Ethernet were XNS (by Xerox), DECNET (Phase III and IV by Digital Equipment) and the TCP/IP protocol. All local area network support for DEC computers either within their proprietary operating system VMS or their UNIX implementation is based on Ethernet.

The protocol stack used by and supplied with almost every UNIX implementation is TCP/IP, a protocol suite developed and standardized by the US Department of Defense (DoD, published as MIL-Specs in 1983) [5]. Most TCP/IP implementations, in turn, are based on Ethernet. In the last few years, all PC network protocol stacks such as NETBIOS or IPX/SPX from Novell are also available on Ethernet.

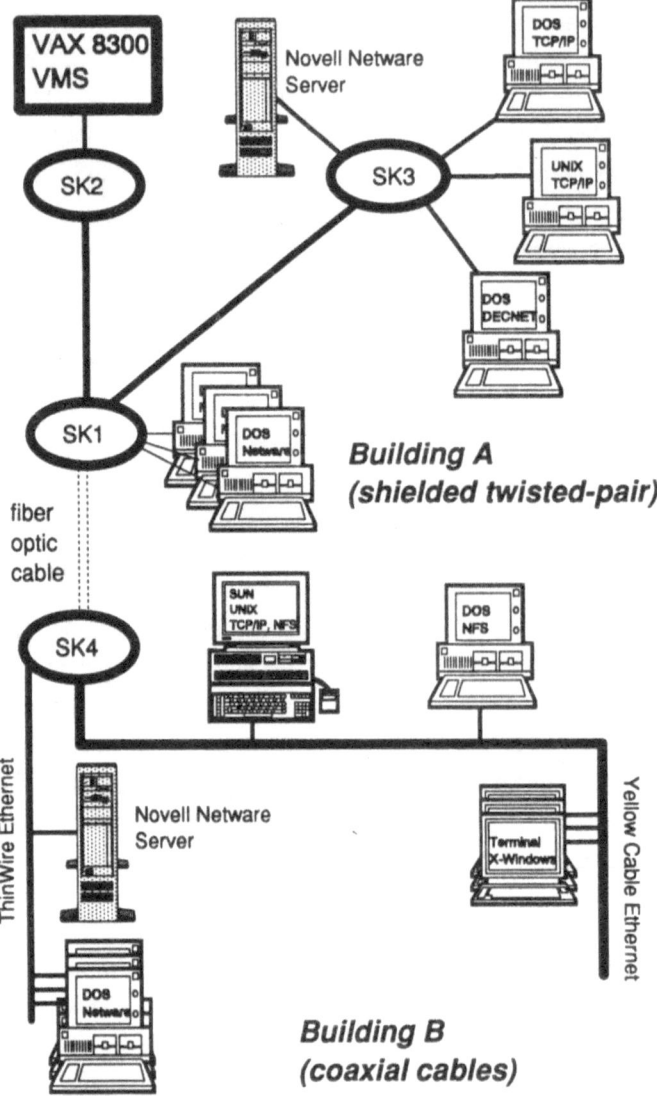

Fig. 2: Heterogenous Ethernet implementation with different protocols (DECNET, TCP/IP, Netware's IPX) using four different cabling media ("Yellow cable" Ethernet, Cheapernet (Thinwire Ethernet), twisted pair and fiber optic).

Ethernet adapters and protocols are supported by nearly every computer manufacturer and supplier of operating systems. IBM now supports its own TCP/IP implementations on

Ethernet in nearly all of its supported operating systems (MVS, VM, AIX, OS/2 and DOS) and within their own hardware architectures (/370, RS/6000, PS/2 and PC). In the field of PC environments, all major suppliers of PC networks such as Novell (Netware), Microsoft, 3COM (LAN Manager, 3Share), IBM (LAN Server) and Banyon (Vines) support Ethernet as a local area network. But whereas DECNET (Phase IV) and TCP/IP, which are based on Ethernet V 2.0, most PC networks conform to IEEE 802.3 (some also to DIX V 2.0) As can be seen in fig. 2, an Ethernet infrastructure (layer 1) can transport quite a lot of different networking protocols used by several operating systems on many different hardware platforms. A true integration of this multi-dimensional heterogenity can currently only be achieved by applications based on the TCP/IP protocol suite.

TOKEN-RING

Token-Ring is a local area network which was developed by IBM in Zurich at the beginning of the '80s and announced as a product in 1985. This implementation was the basis of standardization of the token-ring protocol, which led to the standards IEEE 802.5 and ISO 8802-5 [3].
Access to the shared ring is controlled by an electronic token circulating on the ring. The access protocol and baseband signaling are implemented by a network adapter in each attaching device. Up to 260 devices can be supported by a single ring. Ring speed is is either 4 or 16 Mbps; the actual speed of data transmission depends on a number of factors, including the amount of processing required.

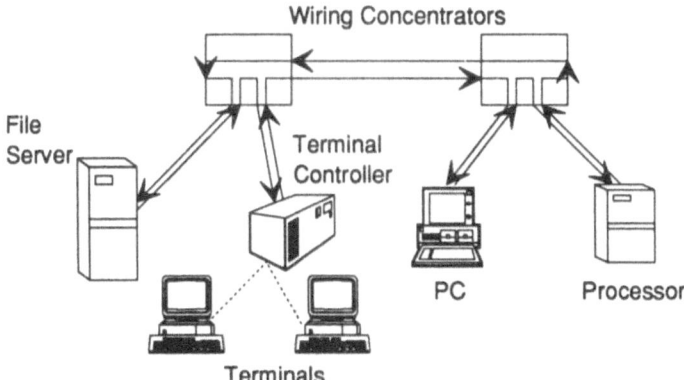

Fig. 3: Star-wiring topology of a Token-Ring network. All attached devices are wired to a central wiring concentrator (sometimes also called MAU = multi access unit). Several of these MAUs can be cascaded in a single or multiple wiring closets.

The physical base of a token-ring network is a star-ring wiring configuration (fig. 3), which combines star wiring with unidirectional, point-to-point, ring-signal propagation. The wiring media of this star is twisted-pair copper wire (for inhouse rings) and optical fibers (for large rings or campus backbones). The copper wire can be either shielded twisted-pair (IBM's type 1) cable or unshielded (voice-grade) twisted-pair cable (IBM's type 3 or AT&T PDS). The former has better electrical properties (smaller signal attenuation and less near-end crosstalk), which result in higher cable lobe lengths (larger ring diameter) and more token-ring stations, but has somewhat cumbersome mechanical dimensions (diameter, bending radius and connectors), which sometimes make cabling difficult and expensive. The typical distance of an network adapter in a 4 (16) Mbps token-ring from a wiring closet is 100 (60) m for shielded twisted pair. Complex campus-wide token-rings can cover an area of a few kms in diameter using fiber optic repeaters. These networks then often have a backbone topology (fig. 4), with several inhouse rings (with copper wiring) connected via LAN bridges to a single fiber-optic backbone ring.

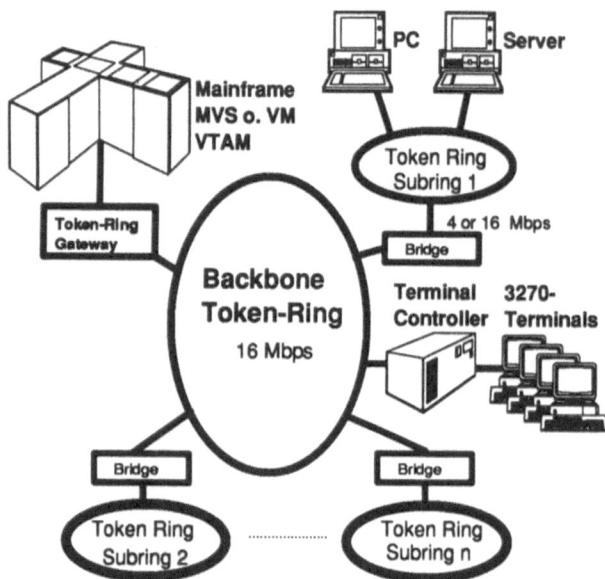

Fig. 4: Typical Token-Ring backbone configuration in a corporate network with IBM host based IS infrastructure.

Very early all major PC-network suppliers supported the token-ring protocol and ported their higher level protocol stacks onto the token-ring such as Novell Netware (IPX/SPX), IBM LAN Server, Microsoft/3COM LAN Manager (NETBIOS), Banyon/Vines and others. The most import argument for a token-ring installation is the fact, that it is IBM's first choice for implementations of their SNA protocols (LU0, LU2 and LU6.2) and products (adaptors in 3174 and 37xx controllers, PS/2). As a result, companies with an existing mainframe infrastructure and corporate SNA net almost always have a token-ring installation.

Access to a token-ring network is supported under many operating systems. All major IBM host operating systems such as MVS and VM (via VTAM, NCP), OS/400 and AIX support it with SNA protocols. IBM implements TCP/IP protocols on the token-ring in most of their OS (VM,MVS/XA, AIX, OS/2) , whereas non-IBM UNIX implementations generally don't have direct access to token-ring networks. OS/2 offers a wide SNA functionality (Extended Edition Communication Manager) on the token-ring and support for a variety of PC network server and requester products. For DOS, there is a wide range for token-ring connectivity products from PC-(IBM-)Host communication and PC network requesters. Having the mainframe and the PCs with DOS or OS/2 on the same (logical) network (fig. ffff) makes PC-host connection easy. The available functionality is a 3270 terminal emulation with file transfer, graphic capabilities (optional) and host printer support. Applications using the SNA peer-to-peer communication protocol (LU6.2) are fully supported.

FDDI

FDDI (Fiber Distributed Data Interface) is the next generation local area network. Under development for more than seven years in the X3T9.5 comittee of American National Standards Institute (ANSI) , FDDI is finally solidifying into a standard [4]. All major computer manufacturers, such as IBM, DEC and others, all suppliers of high speed networking equipment have announced full support of this new architecture and have already presented several products.

The FDDI network is based on dual counter-rotating 100 Mbps token-rings with fiber optic technology. The token passing algorithm used by FDDI-compliant networks is similar to IEEE 802.5 (token-ring), but with some significant enhancements in the areas of bandwidth allocation, fault tolerance and topology. The number of stations in a FDDI network is limited to 500, with a maximum of 2 kms between two adjacent stations. The FDDI standard comprises the OSI layers 1 and 2a (Physical Layer and Media Access Control) which are split into four entities: PMD, PHY, MAC and SMT (see fig. 5). At least three of these four standards are already approved by the X3T9 committee.

A FDDI network offers three main advantages over existing network architectures and structures, namely its increased bandwidth, its acceptance as a new and supported standard by all major computer and network equipment and the capability to span a wider area. There are several important consequences to draw from these points.

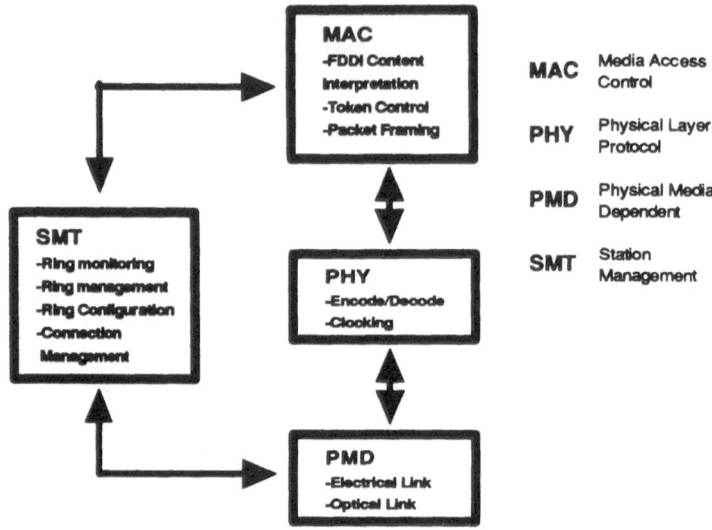

Fig. 5: Internal structure of FDDI network node.

First of all, FDDI obviously is an ideal candidate for a campus-wide LAN backbone structure (see fig. 6). It offers direct integration of important ressources as hosts, corporate database or backup servers ("optical jukeboxes") as single or dual-attachment stations. Even more important, existing Ethernet or Token-Ring networks can be connected via LAN-to-FDDI bridges (or (b)routers). Besides building new corporate LANs based directly on FDDI, this offers a relatively simple,"soft" migration strategy from conventional backbone LANs (either Token-Ring or Ethernet, see figs. 2 and 4) to a FDDI backbone. This means that hundreds (or thousands) of low-cost, (relatively) low-performance PCs, where network performance is limited by their CPU's handling of the protocol stack and not by the physical bandwidth of the network can be attached or stay attached to their conventional LAN, but still benefit from the increased speed of the backbone. Second, where network speed is a bottleneck as in some real-time imaging or large database file transfer applications, a direct FDDI access can resolve these problems. Third, connecting computers of manufacturers favouring different network architectures, say an IBM MVS/XA host connected to a Token-Ring with a DEC VMS mini attached to an Ethernet, is a cumbersome and expensive business especially if you want a bidirectional, high-functionality connection. Beside special gateway software you need some (special) piece of hardware to bridge the gap from the /370 channel, an SDLC link or the Token-Ring to the Ethernet. Clearly an FDDI connection of both machines does not solve the whole problem, but supplies at least the bottom two OSI layers of the solution with full vendor support.

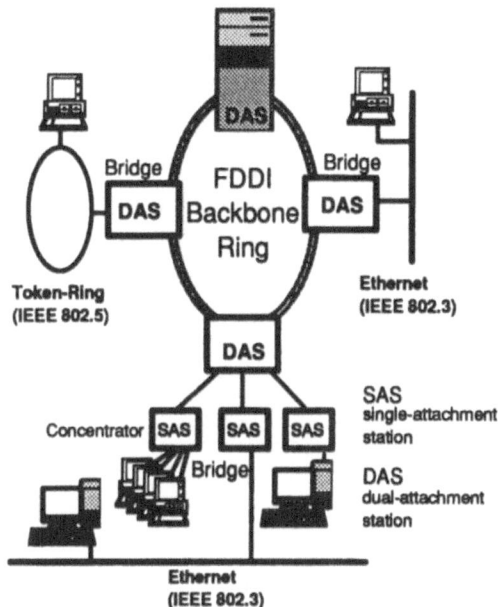

Fig. 6: Structure of a FDDI backbone network. Workstations can attach either directly as single or dual-attachment station or via bridges at "conventional" LANs.

FUNCTIONALITY

For end users of computers, technical aspects of the underlying network are generally of little interest. Their focus is on high functionality and ease-of-use. The ideal situation, from this point of view, would be a fully transparent distribution of all needed resources, such as file services, printer support, data backup, electronic mailing, database access and computing power. Currently, only a few of these issues can be solved the way an end user might dream of. Transparent file services is one example where this is almost true, at least in homogeneous operating system environments. In most other cases the end user does not get what he or she wants, but what is available as a product. Some of these basic services offered from any environment to nearly any other are terminal emulation, file transfer (binary files or text files), remote printing and remote job entry. All major computer manufacturers and even third party software developers offer products to integrate PCs and workstations connected to local area network into different networking schemes. The most prominent examples are programs which offer 3270 terminal emulation, file transfer and host printer support from a PC to an SNA host via token-ring, a DECNET end node functionality or basic and advanced TCP/IP features as telnet, ftp up to NFS clients on a PC at an Ethernet.

The next step in networking software, which is already taken in some widespread concepts such as NFS or X-Windows, is a general and consequent application of program-to-program communications or client-server architectures, such as IBM's APPC protocol LU 6.2 or concepts like the remote procedure calls (RPCs). In such an environment, a program on a machine, the client machine, converts end user actions or enquiries to arguments in procedure calls, that are transmitted via a LAN to a remote machine. There another program, the server, receives the parameters and executes the requested action, a database transaction, a disk sector read or an involved mathematical computation. The results of this computation then is sent back to the client program, which returns the results to the end user, eventually transformed to a different form, such as a graphics image or a spreadsheet.

[1]   Klaus H. Stöttinger, "Das OSI-Referenzmodell", (Datacom, Pulheim, 1989)

[2]   R. Marquardt et al., "Ethernet-Handbuch", (Datacom, Pulheim, 1987)

[3]   IBM, "IBM Token-Ring Network Technology", (IBM, Document: GA27-3732-0, 1986)

[4]   Data Communications, "Getting a handle on FDDI", (McGraw-Hill, LAN Strategies, Special Issue, June 89)

[5]   Gerhard M. Glaser et al., TCP/IP, (Datacom, Pulheim 1990)

Public Broadband Networks –
Present State and Future Perspectives

Paul Christ

Communications Systems and BelWü-Coordination, Computing Center
University of Stuttgart, Allmandring 30, W-7000 Stuttgart 80, FRG

Keywords: Large-scale simulations, distributed supercomputing, Stuttgart Tetrahedron, Forerunner Broadband Network (VBN), MAN, DQDB, Frame Relay, B-ISDN, STM, ATM, SDH, SONET, Gigabit Network Testbeds

Abstract: Large-scale technical-scientific Simulations in Supercomputer-Network-complexes, exemplified here by the 'StuttgartTetrahedron', represent - among others - one of the scenarios to be extended to a national and eventually European scale in order to guarantee the competitiveness of the continent in the long run. Starting from the Stuttgart experiences the present situation and future alternatives for public broadband networks as a prerequisite for such a geographic extension are investigated.

INTRODUCTION

The University of Stuttgart Computing Center 'RUS' runs one of the most advanced Supercomputer-Network-complexes in the world, comprising presently a Cray 2 super computer, medium to high-speed networks ranging from the 10Mbit/s Ethernet to the Gigabit/s UltraNetwork, [30], and a visualization laboratory. The RUS compares well with the US NSF supercomputer centers like NCSA or, in Europe, ONERA the supercomputer center of the French aeronautics industry. Large-scale simulations in chemistry, fluid dynamics, combustion engineering, or climatology are some of the typical applications dealt with in such centers.

It is widely recognized such applications to be a key factor for future competitiveness of European industry and science.

Of critical importance in the field is always knowledge. Collaboration between scientists supported by high-speed networking is one means to develop such knowledge necessarily across borders in Europe.

Given the 'ideal' campus situation where supercomputers, Gigabit/s networks, fast fileservers, and visualization equipment are often at hand - what are at present the possibilities - within Germany and Europe - to extend the campus by fast networks to a national or even continent wide scale? Which possibilities and alternatives in public broadband networks are forseeable in the future?

U. Harms (Ed.)
Supercomputer and Chemistry 2
© Springer-Verlag Berlin Heidelberg 1991

THE STUTTGART TETRAHEDRON

From a conceptual point of view the Stuttgart supercomputer-center is known as the 'Stuttgart Tetrahedron' according to figure 1.

In the present context mainly two points are of importance: First, all corners of the Tetrahedron are based or will be based very soon on Unix and related standards. Second, emphasis in supercomputing in the 90ties will be on the 'storage' corner of the Tetrahedron and the flow of data between fast filservers, supercomputers and workstations. This in turn will increase the importance of broadband networks both locally and in the wide area.

As an example the following Figure 2 shows the typical flow of data in a fluid flow simulation-visualization.

Within such scenarios tasks are distributed between supercomputer and workstation by the locally developed 'DFN RPC' optimized for large data transfers, [24], [29], (see figure 3). The choice of one of the three ways of distribution is possibly of great influence on the amount of data to transport in such application scenarios.

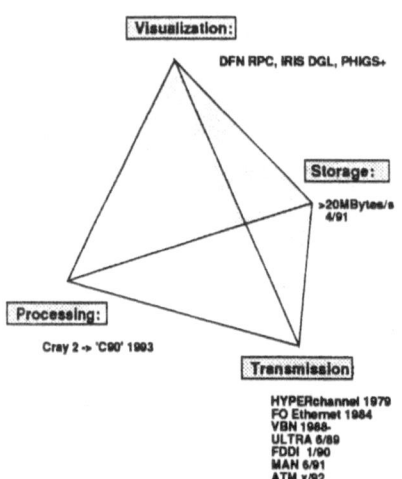

Figure 1 : The Stuttgart Tetrahedron

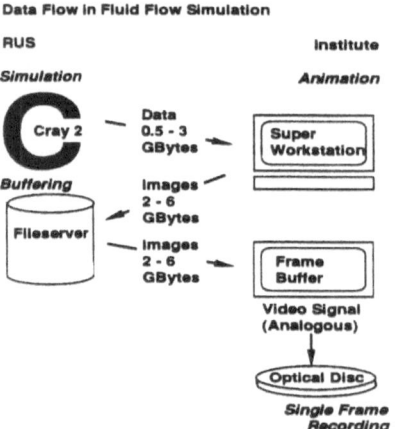

So far the Stuttgart Tetrahedron is extended into the wide area by networks like BelWü, 'Baden-Württemberg's extended LAN', [13], DFN WIN, the X.25 science network in Germany and recently IXI , the COSINE European 'internatinal X.25 interconnection' and a WIN-Internet link to the US. The maximum transmission rate here is in most cases only 64Kbit/s.

In the field of high-speed wide area networking Stuttgart has carried out projects within the 140Mbit/s VBN; a 34Mbit/s MAN trial is in the planning state. Both projects are described below.

Figure 2 : Data Flow in a Fluid Flow Simulation

THE FORERUNNER BRODBAND NETWORK VBN

This chapter describes the VBN, a 140Mbit/s circuit switching pilot-network of the Deutsche Bundespost Telekom, the Stuttgart VBN trials and their relevance in the context of high-speed networking for research in Europe.

The Vorläufer Breitbandnetz VBN

The Forerunner Broadband Network, VBN (Vorläufer Breitbandnetz) is a follow-up pilotproject of several videophony trials the Deutsche Bundespost (Telekom) has carried out around 1985. The main application the VBN was aiming at was accordingly videoconferencing. The first trials within the VBN have been started in 1988, the formal opening of the service was in February 1989.

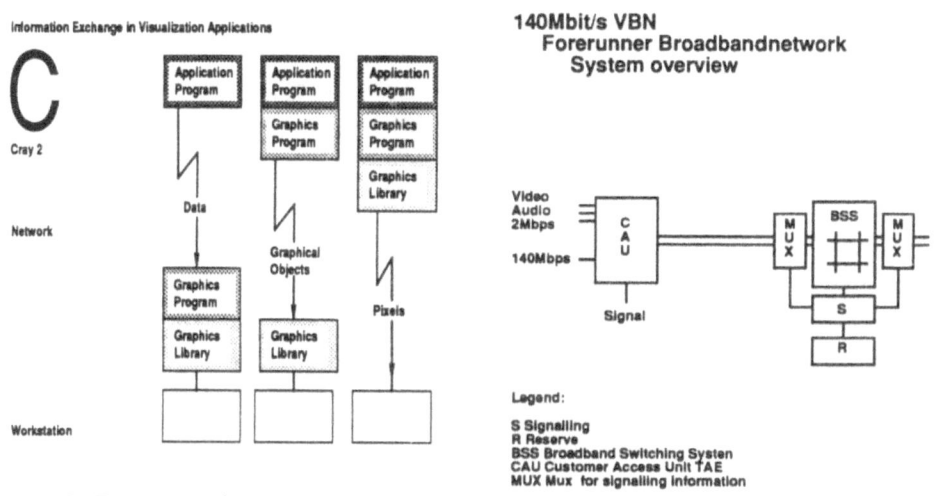

Figure 3 : Three ways of distributing an application by Remote Procedure Calls

Figure 4: VBN system components

The VBN designed for 1000 has presently about 350 customers and shows a high growths rate.

Within the VBN 2-party conferences can be established by the customer, dialing manually the second partner; n-party conferences can be arranged by a reservation center located in Cologne. The main components of the VBN are the customer access unit (CAU) and the switching center. In addition to the video/audio-interface set the CAU provides a 'raw' data interfaceof about 140Mbit/s (see figure 4).

The VBN, using mainly 140Mbit/s channels out of the present fiber-optics transmission infrastructure, the so-called 'overlay network' of the Telekom, covers (West-)Germany as a two-level network according to figure 5.

In addition to video/audio-oriented pilots, like MEDKOM in medicine, also 'data'-oriented projects have been initiated within the VBN. Accessing the VBN via the 140Mbit/s interface of the CAU the Gruner and Jahr publishing company for example is able to print the 'SPIEGEL' magazine both in Hamburg and Nuremberg.

The University of Stuttgart has started VBN projects in summer 1987. First results have been reported at the first RARE-CEC Symposium on High Speed Networking in 1989, [13]. Since then progress has been made according to figure 6: The VBN-based full-speed Ethernet includes now Kaiserslautern, Karlsruhe, Tübingen and the two Stuttgart campuses.

VBN BACKBONE NETWORK

MAIN SWITCHING CENTERS

REGIONAL SWITCHING CENTERS

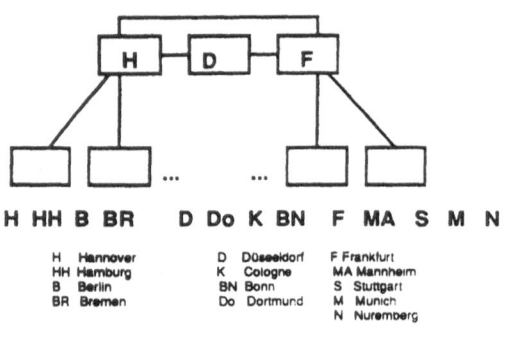

H HH B BR   D Do K BN F MA S M N

H Hannover	D Düsseldorf	F Frankfurt
HH Hamburg	K Cologne	MA Mannheim
B Berlin	BN Bonn	S Stuttgart
BR Bremen	Do Dortmund	M Munich
		N Nuremberg

Figure 5: The VBN backbone

Figure 6 shows another interesting application of the VBN-to-Ethernet equipment demonstrated at the symposium for computer assisted radiology, the CAR '89 June 1989, [26], in Berlin: For the demonstration of remote computer tomography an Ethernetconnection was established from a hospital to the ICC congress center Berlin - using the VBN-oriented network adapter within the BERKOM B-ISDN testbed, [14].

In addition to the VBN-to-Ethernet equipment a first version of a multiplexer VBN-to-FDDI, F-Mux for short, has been completed. The F-Mux provides only for the transport of the 4b/5b encoded 100Mbit/s FDDI physical data stream over the VBN - but is not related to the FDDI MAC. The FDDI-PHY, still wasting some of the VBN bandwidth, FDDI was choosen because it is a standard.

The first application of the F-Mux was an UltraNet-To-UltraNet connection between Stuttgart and the CeBit '90 fair in Hannover. From the Stuttgart Cray 2 to an Ultra frame buffer in Hannover more than 95Mbit/s out of 100Mbit/s possible were achieved. This is probably still a world record for the biggest distance*bandwidth product. The same technique is used to connect the Alfred-Wegener-Institute in Bremerhaven which runs large simulations in climatology on the Cray 2 in Stuttgart.

The position of the F-Mux between UltraNet and the VBN is shown in figure 7. The challenge in the F-MUX design lies mainly in regaining and controlling the clock at the receiving side, [12]. The present F-Mux version adjusts the receiver clock with a phase-locked loop by observing the half-full occupancy of a receiver buffer. A second version will implement here some improvements. Experiments to extend real FDDI-rings over the VBN have been started, a FDDI-FDDI remote brigdge is under development. The German Research Network, DFN (Deutsches Forschungsnetz), has accepted a project in order to make systematic use of the VBN dialing and switching capabilities.

VBN Political Situation, Tariffs And Future

Within the evolving new structure of the Bundespost responsibility for the VBN is on the 'telephone' i.e. on the monopoly side of the Telekom. Tariffs within the VBN exist only for videoconferences, reaching DM 600 per hour for distances greater than 100 km. In general the VBN is not allowed to compete for 'data' with the value added service side of the Telekom. Data-oriented projects within the VBN therefore have to adhere to some principles; they have to be based on dedicated contracts and have to be 'new and interesting' in order to justify special tariff rates. Those conditions make the VBN probably not suited for a general data communications infrastructure. Nevertheless the VBN remains forsome years a very interesting possibility for advanced high-speed networking projects within Germany.

Finally Telekom plans should be mentioned to provide videoconferencing on the basis of CCITT H.261 and 2Mbit/s lines within the VBN.

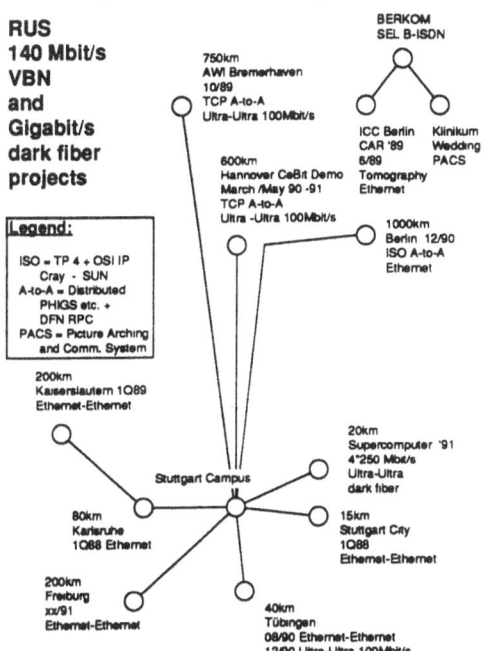

Figure 6: Stuttgart 140Mbit/s VBN and Gigabit/s Network Trials.

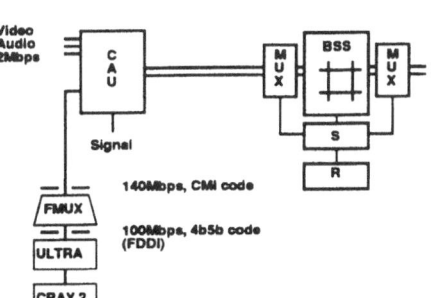

European Relevance Of The VBN Trials

Within the given context the following points can be learned from the Stuttgart VBN experiments. First, it is relativly easy to interface to any preesent high-speed infrastructure up to 140Mbit/s and even higher. Interfacing to SDH/ATM (see below) structures up to a Gigabit/s could be learned from colleagues in the five 'Gigabit Network Testbeds', which have been started as a cooperative effort of industry, science and carriers in the USA, [21], [33], [12].

Figure 7: Position of the F-Mux betweeen UltraNet and the VBN

Second, given a well functioning 'rich' infrastructure including UltraNet Nodes and other of the shelf switching equipment it is possible to establish high-speed pilot production networks complementing exisiting production infrastructures like the German X.25 WIN of the DFN or the European IXI.

Third, Telecoms might be willing to support 'interesting' pilot projects - even on a two or more country basis - which do not interfere with tariff dynamics in general.

Fourth, as always in the short history of local and wide area high-speed networking not the technique but relevant end-user applications finally are the enabling factor within the complicated techno-political environment of public broadband networks.

In the end, the Gigabit Network Testbeds and the Stuttgart VBN trials are 'interesting' in that both are leading edge 'orthogonal' cooperations - including transmission systems aspects, network scenarios and advanced supercomputer applications - i.e. covering OSI layer one to seven

THE STUTTGART MAN FIELD TRIAL

'MANs' - Metropolitan Area Networks - are 'LANs with a media access technology insensitive to greater distances'. The prevalent MAN at present is the DQDB system - distributed queue dual bus - standardized by the IEEE 802 committee. This committee is well known for its work in standardizing Ethernet and the IBM token ring as items 802.3 and 802.5 respectively. DQDB is dealt with as 802.6, [20].

Within this context SMDS - switched multi-megabit data services - is a concept developed by Bellcore which aims at LAN interconnection on DQDB basis as a public service, [4].

The technical challenge of DQDB in contrast to other LAN/MAN techniques will be explained in the second part of this paper.

In the following section we are now going to describe the Stuttgart MAN trial.

The MAN Trial

The University of Stuttgart computing center will participate in one of the two evolving MAN field trials of the Telekom in Germany - the second taking place in Munich .

Starting in the third quarter 1991 the trials will have three phases: A one years 'systems testphase' - free of charge -, a two years pilot production phase and a two years optional phase. The tariffs for the production phase so fare are only 'not to exeed' and will be adjusted to the development of the general tariff structure. The equipment will be provided by Alcatel/SEL and Siemens respectively on the basis of QPSX development, [2], [1], [31].

The services foreseen will be mainly Ethernet-to-Ethernet connections. For this purpose the 'customer gateway' CGW is going to provide an Ethernet bridge or a router. In case of the Stuttgart trial the bridge solution will be used - the border of the customer networks are then represented by private routers in order to provide for additional private network management and routing functions - security being the most important one.

The backbone will run at 34 Mbit/s. Figure 8 shows the proposed trial configuration. Within this configuration two 'virtual private networks' VPNs will be established separating one industrial participant from the university and its partner, the scond participant from industry.

The university VPN and the second one will carry traffic mainly from large scale simulation on the Cray 2 and CAD/CAM applications respectively.

MAN Political Situation, Tariffs And Future

Within the evolving competitive environment for data services the MAN trials represent one of the more advanced and innovative actions of the DBP Telekom in order to test the market for LAN-to-LAN interconnection. DQDB will have and has already to face other alternatives as competitors in this field among them Frame Relay being probably the most noticeable (see below). In order to be attractive the MAN service has to fulfill certain conditions the most prevalent in the field of 'universal applicability' and 'costs'. Beginning with costs, a service provider has to keep in mind that for potential customers running routers at large scale in their extended LANs the sheere switching capability of a MAN is of minor interest as long as this capability is not yet accompanied by wide area coverage and market penetration. This means MANs have to competewith leased lines! 'Universalapplicability' means MAN have to have the potential to

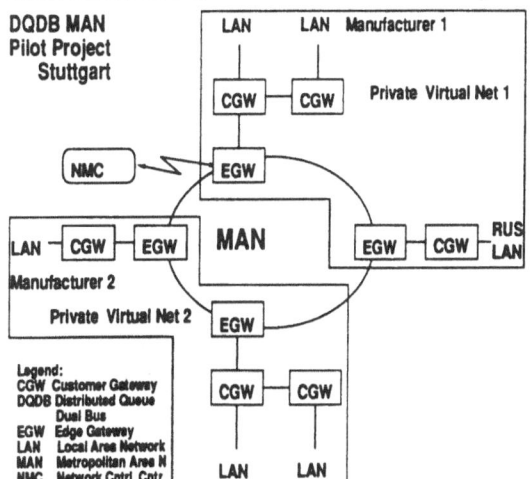

Figure 8 : The Stuttgart MAN trial

become 'the' (main) network of 'the company' - at least on national scale; i.e. on the one side, in addition to TCP/IP workstation traffic MANs have to carry the unavoidable SNA traffic of 'the company' - on the other side, MANs have to be 'sensitive' to the problem of the inner-company telephone traffic and provide here a perspective for solutions.

The future of MANs is generally considered as becoming one type of the local networks required for the distribution of the services of general 'B-ISDN' (see below).

The European Relevance Of The Stuttgart MAN Trials

Despite differences in national policies concerning the MAN versus real ATM deployment a DQDB-MAN does open an European perspective for high-speed networking:

First, DQDB products will be available from European manufacturers which, at the same time, make those products wihin ETSI European standards, [1].

Second, DQDB is closely related to the broad European ATM activities and fosters hard- and software development necessary to exploit cell based communication systems, [5], [12], [21], [14], [17].

Third, DQDB has the potential to integrate classical data and real-time multimedia communications, [7], [16], [35], [22], [23], [3].

PUBLIC BRODBAND NETWORKS - STATUS AND PERSPECTIVES

Broadband-ISDN

Following the description of the Stuttgart activities in two special brodband networks - namely VBN and DQDB-MANs - in the second part of the paper we are going to give an overview over some concepts pertinent to public brodband networks in general.

The first concept to be mentioned here is that of an 'ISDN', i.e. the integration of 'all possible services' within one universal digital network. A typical strategical chart expressing this dream is given in figure 9.

As the ultimate step the integration into an IBCN - the Integrated Broadband Communication Network - is shown. The IBCN eventually integrates 'dialog' and 'distribution' services - named after the capability of the receiver side of the service scenario to answer or react to the sender. Telephony and video distribution are the most prominent examples for such services.

As a direct consequence of the integration of such different services within the same network the need for the provision of channels of rather different bit rates at the users network interface was evident from the very beginning.

It was also obvious that video - be it dialogous or distributive - would require high bandwidth. The term 'Broadband ISDN' then was introduced in order to designate both the broad service spectrum and the high-speed orientation. ('Broadband' within this context must not be confused with the old frequence multiplex technique widely used on coax cable systems) .

At the same time this term should emphasize the evolution path from the narrow band ISDN.

Circuit-Switching, STM - B-ISDN

Historically the first architectural approach to B-ISDN was 'Broadband B-ISDN is nothing but a fast narrow band ISDN'. What does that mean? To the fixed set of two 64 Kbit/s and one 16Kbit/s channel of

ISDN for data and control respectively B-ISDN adds two other fixed data channels of 2Mbit/s and about 140Mbit/s. At the network interface then the end user, i.e. the 'terminal' does 'see' this fixed channel structure. In this architectural concept for a B-ISDN the term 'integrated' applies first to the access

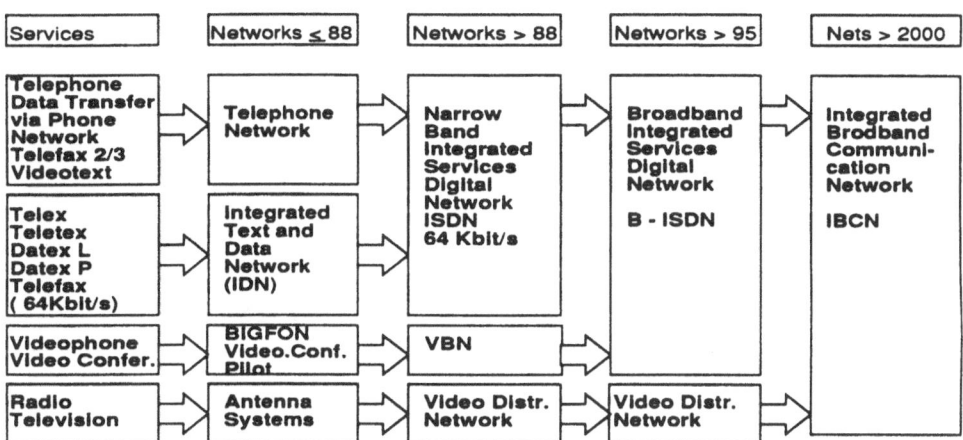

Integration of Public Service Networks

Figure 9 : Global network Evolution

integration - all services are accessible via the one and the same 'wire' which is nowadays conceived presumably a glas fiber. But also beyond the terminals access line the term 'integrated' does still apply - there is only one network carrying all the different services. The concept of such a B-ISDN was investigated till the second half of the eighties, see [15]. Under the name of BERKOM the Deutsche Bundespost - now DBP Telekom - has initiated a B-ISDN field trial in Berlin, [14].

But in 1988 the CCITT as the relevant body has thoroughly revised this B-ISDN concept. The major three problems with the initial B-ISDN concept can be summarized as follows:

- lack of flexibility: As future user needs and market reqirements are widely unknown today, a fi - xed channel structure may be not appropriate.

- technical complexity: Despite the 'inegrated network' concept switching n fixed channels of different speeds practically implies n different switching structures, i.e. virtually n different networks.

- lack of access line multiplexing: If the 'terminal' becomes an intelligent device with storage and processing capabilities the need for more 'sessions' over the access line at the same time arises: If the terminal is a modern workstation, more than one program is likely to communicate at the same time with partners at different places. This argument applies even more if the 'terminal' is a whole private network of workstations.

One may notice that the third point 'multiplexing' can be taken only gradually as there is a multiplexing of the access line into the fixed channels in the old B-ISDN concept as mentioned. The technique to accomplish this type of multiplexing is called Time Division Multiplexing. In this technique a fixed frame is periodically sent over the medium. A channel then is represented by a fixed multiple of a basic number of bits at a fixed position within the frame - which appear therefore always at the same time after the start of a new time period. In the B-ISDN context this has led to the term 'Synchronous Transfer Modus' (STM). A sender sending via a certain channel has to fill its data bits in fixed quantums into the appropriate frame position; a switch then in turn identifies the data of a channel by their position within the frame and switches them along routes or circuits, pre-established at call set-up time by a separate 'signalling' system.

If the sender has nothing to send the channel capacity remains unused; but, at the same time, the system guarantees always a constant bandwidth and delay characteristics - namely those of that specific fixed channel. These characteristics in turn are of key importance for real-time applications like video and voice.

Networks with such a structure were historically called circuit switching networks. In conclusion we see in the established terminology that the initial B-ISDN was a circuit-switching STM network.

ATM B-ISDN

The mentioned lack of flexibility of the STM B-ISDN can be overcome by a seemingly simple change of the way transfering user data within the transmission frame: Adding the overhead of a - small - address in front of the data, the data can be placed then everywhere within the frame. Data belonging to a certain cannel therefore may appear at different points in time - i.e. 'asynchronously' - with respect to the start of the frame. This has led to the term of 'Asynchronous Transfer Mode' for this technique. Using this terminology we understand the saying 'In 1988 the CCITT has decided ATM to become the technology of B-ISDN', [10].

The reclaimed flexibility results as follows: First, there is no longer a need for empty positions in the frame, in the case of one sender having nothing to send another sender can forward its data. Second, by skipping an appropriate number of frames any effective bitrate within the limits of the line capacity can be achieved without wasting bandwidth. Third, addresses attached to the data provide for an effective mean to multiplex the single access line from the users 'terminal' into the network. A virtually unlimited number of 'virtual' channels - as opposed to the few fixed 'physical' channels in the case of the STM B-ISDN - can be established and coexist simultaneously in time.

In principle ATM could be based on variable-length 'cells' - as the ATM data units are called - or fixed-legth cells. For reasons of technical simpilicity and effectiveness the process of standardization has led to the compromise of a 53 bytes fixed-sized ATM cell.

Are there only advantages with the ATM technique?

First, despite of the possibility of an 'unlimited' number of virtual channels - the total capacity of the access line, of course, is available only once. If one divides this capacity between a number, for example, video channels the quality of them may suffer. Second, a major challenge in an ATM network is to guarantee a certain average troughput and the establishment of upper bounds for delay and jitter for applications like voice and video which are very sensitive to such characteristics of the data stream. Third, there is also criticism of ATM from the pure data communications camp: The small fixed size ATM cells are not 'natural' for data communications; computers are not in favor of many interrupts caused by small network data units.

It is fair to say that there are still a lot of open research problems with ATM networks. But, despite of the potential problems with ATM a lot of companies and telecoms are readying ATM field trials. Within the the BERKOM testbed after the initial installation of two B-ISDN STM switches from SEL and PKI respectively there is now also an experimental Siemens ATM switch, [14]. This latter fact nicely illustrates the shift in paradigms from STM towards ATM in the B-ISDN context.

One will have noticed that the ATM technique is covered by the general concept of packet switching. In the next section we are therefore going to position ATM into this framework.

Variants Of Packet Switching

Classical Packet Switching: Figure 10 repeats the basic principle of packet switching: The 'terminal' attaches addresses to packetized data, the network then - which has to have intelligence - moves and switches data according to their addresses, the access line and internal transmission lines can be statistically multiplexed.

In the ISO OSI modell a layered concept of data communications does apply. In our context OSI states that 'networking' is a function of the Layer 3 which therefore bears the name of 'the netwok layer'. All widely known packet switching network architectures - like DECnet, TCP/IP internets, X.25 and SNA networks - adhere basically to this OSI scheme. Coincidently layer 3 data units are usually called 'packets' - therefore: OSI networks do switch packets . The question then arises, are there other possibilities for 'networking', and, on which OSI layer are the ATM cells to be positioned?

Frame Relay: In the OSI modell network packets are carried safely between pairs of nodes of the network over point-to-point links. Confined to travel only from A to B over a fixed 'wire' layer 2 data units - called 'frames' - providing for the safe transport of network packets usually do not need real adresses. Packets then arriving at network nodes are unpacked, and, after being repacked into a new frame, put onto the next link according to their network address. It was only recently that the CCITT in search of a packet switching service within the circuit switching narrow band ISDN in its Study Period 1984-1988 has conceived the idea of such a service based on the switching of frames instead of packets: Why not place a meaningful network adress into frames, retain indiviual frames trough the network and avoid the processing overhead of layer 3 handling within the network nodes, [8]. But as one immediately sees, this technique, best known as Frame Relay, is not without problems, [18]: Error control has now to be carried out end-to-end between endsystems (It has been shown that such a scheme can do equal or better than hop-by-hop error control!, [36]). The hard part to do is what is called congestion control - how to avoid the network nodes to become overloaded and massively drop packet as a consequence.

Packet Switching

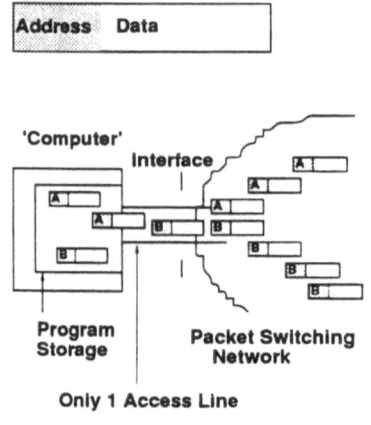

Figure 10: Packet switching priciple

In the meantime Frame Relay has gained a lot of momentum and support from vendors like StrataCom, DEC, CISCO and other parties like Northern Telecom or US Sprint, and is about to become an ANSI T1S1 standard, [18], [34], [25]. In the scientific community IP over FR is dealt with by the mailing list 'IPLPDN' (IP over large public data networks). Specified up to but probably not confined to 2 Mbit/s FR is certainly attractive as a short to medium term solution for private network providers aiming at IP, SNA, DECnet and X.25 connectivity. Although a non-goal of FR voice is often provided by FR vendors on the same multiplexer or switch and transmission infrastructure.

<u>ATM Cell Switching, DQDB Cell based media access:</u> To position ATM and DQDB finally at the right OSI layer one has to see that in an ATM network layer 2 frames are segmented into fixe sized ATM cells which are finally reassembled at the destination into frames again. In the OSI modell this places ATM clearly one layer down below the frame layer, i.e. to layer one, the Physical Layer. This goes well along with the fact that ATM cells do nicely fit into the layer 1 transmission - 'frame' (this 'frame' was named by the tranmission people and must not be confused with the OSI layer 2 frame). On the other hand, as we have seen, ATM cells carry a 'little' address - called a 'label' - and are moved and switched throughout the network according to their labels - which is in OSI terms a network, i.e. layer 3 functionality.

DQDB, as a network based upon a shared medium, - in order to achieve the delay and throuput characteristics required for voice and video accesses its 'wires' on the basis of cells too (see figure 11).

The following figure 12 shows the switched protocol data units and their relation in classical packet switching, in Frame Relay and in ATM respectively.

Transmission And Switching Technology

<u>SDH and SONET</u>: Before guessing about the future of public broadband networks one has to mention two major technical developments - the one actually going on the other imminent at most but equally important; the first development to be dealt with is the introduction of a new transmission infrastructure called Synchronous Digital Hierarchy, SDH, [9] in CCITT terms. SDH was invented in the USA by Bellcore under the name of SONET - Synchronous Optical Network. The main point in our context is not 'optical' but 'synchronous' (the DBP Telekom has under the title 'Overlay Netz' fibers installed for years). 'Synchrounous' stands for flexibility and potential for cost reduction at the lowest level of tran-

mission network provision and management. 'Thinner' transmission pipes can be grouped together into 'fatter' ones or extracted from fat pipes and forwarded into new direction fully electronically under software control in only one processing step. In the existing so called (plesiochronous) PCM systems the same task requires a hierarchy of hardware-wise multipex- and demultiplex-stages.

DQDB

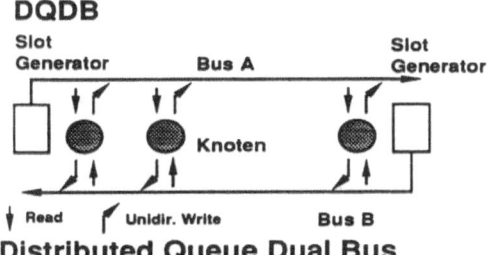

Distributed Queue Dual Bus

- like Ethernet, Tokenring, FDDI -
    Shared Medium, Multiaccess Network

- unlike Ethernet, Tokenring, FDDI
    not Paket/Frame-based-
    but Cell-based - likeATM <-
    Aimed at Voice, Video, 'Data'
    not Distance- and Speed-sensitive

- IEEE 802.6 Metropolitan Area Network MAN Standard

- Basis for Pre-ATM-MAN projects (SMDS)

Under the working title of 'NK 2000' the DBP Telekom has recently started a major effort to introduce such a SDH based infrastructure within Germany, [32].

We should mention that SDH is a basic transmission technology which can be used both for STM and/or ATM based B-ISDN.

Optical Networks: As mentioned before fiber optics based transmission systems are in operation for some years and it would not be easy to overestimate the impact fiber optics already had on virtually all branches of information transmission - ranging from

Figure 11: DQDB Distributed Queue Dual Bus

optical Local Area Networks to intercontinental links. But one has to see that we are only at the beginning of the 'fiber optics revolution'. So far FO systems are used mainly to replace copper lines between conventional electronic equipment - leaving bandwidth, topology etc. of the old systems nearly unchanged. The great potential of FO systems therefore has still to be materialized - this includes for example the introduction of more and more all-optical components like amplifiers or even switches. In our context one of the most important point in the evolution of FO systems is the more effective usage of the fiber itself. So far information transfer is achieved by just switching diodes and lasers on and off. The most attractive alternative today to overcome this pure 'basic band' technique is, among others, WDMA, wavelength division multiple access: Each channel or circuit gets its own 'colour'.

We omit here the problems still to be solve with WDMA and mention that BERKOM the optical links consist of only one fiber carrying light of two colours - as opposed to the VBN links wich always have to have two fibers.

Public Broadband Networks - Perspectives

Which development of Public Broadband Networks can be predicted from the technical discussion given so far - in Germany, Europe, worldwide?

We think the following thoughts apply indeed worldwide.

First, due to the vast investment into existing networks there will be no revolution but always an evolution. Second, it seems rather safe to predict the deployment of SDH infrastructures at large scale - providing for support of the old PCM systems (transition to and from, ecapsulating old in new data streams).

Packet Switching Variants

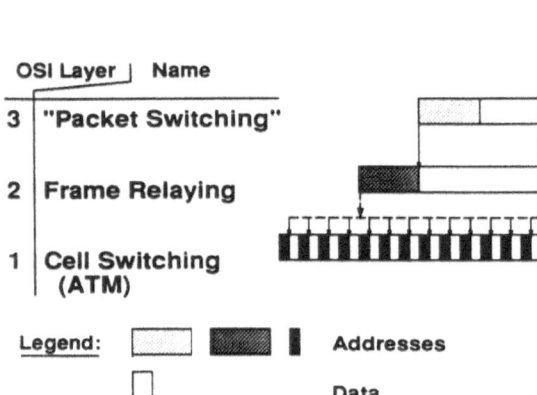

But when will we see B-ISDN in the broad sense? The major factor here seems to be the fact that after the liberalization and de-monopolization the telecoms probably have to look for much shorter return of investment cycles - this may prevent such global concepts like B-ISDN ever to happen at all.

Figure 12: 'Packet Switching', Frame Relay, Cell Switching

Will we see ATM at least as backbone network - for 'classical' packet switching networks and perhaps MANs ? Presently there are two 'world-lines' one can draw: First, Frame Relay networks could speed up private and public packet switching networks up to 2 Mbit/s and more. This can be 'easily' achieved by pure software changes in classical packet switches from CISCO and other players in the private networking scene or even in X.25 switches from Northern Telecom - on which the present X.25 German Research Network (WIN) is based. Such FR networks could very effectively also carry traffic from SNA and DECnet. Due to the recent progress made by the CEC's DG XIII on the application of 'Open Network Provision' to leased lines across Europe (2 Mbit/s G.703 leased lines and a uniform cost based billing scheme across Europe by July 1, 1992) one could imagine pan-European FR networks in the science community very soon. As mentioned above this movement will be accompanied by general criticism of the cell based approach for pure data communication - be it ATM or DQDB. In summary all this could defer the large scale deployment of ATM - until all-optical WDMA networks are just ahead of us, or at least seem to be. Drawing this picture one has to keep in mind that WDMA again is a priory a circuit switching oriented technique as the present TDM is. That this confusing scenario is seen by other parties too can be observed in RACE, the CEC founded 'Research for Advanced Communications in Europe', where now dual ATM and WDMA structures are investigated, [28].

Second, despite of FR, all over Europe, after the US and in Australia, DQDB or pre-DQDB based MAN field trials are scheduled to start soon and have even begun in the UK. LAN interconnection field trials based on SMDS/DQDB MANs orforerunner technology are carried out around the world, [19], [4]. Within the Internet community IP over SMDS was dealt with by a SMDS mailing list. This has resulted in a draft RFC, [27].

The same holds true for ATM field trials within a three years time scale. At the European level SDH infrastructures and ATM network concepts are launched mainly by the DBP Telekom, for example under the names of METRAN and EPAN.

This means, yes, we will see ATM networks; but it is totally open whether this will lead to an univeral B-ISDN - not to mention the inclusion of TV distribution into the same network.

CONCLUSION

We have discussed the extension of the Stuttgart Tetrahedron into the wide area by VBN in the past and MAN techniques in the near future. The concept of B-ISDN based on STM and ATM was described. ATM was related to other - potentially competitive - forms of packet switching. The imminent deployment of SDH transmission infrastructures was made evident. The longterm development of WDMA-optical networks was mentioned. Finally, it was doubted that there will ever be only one universal public broadband network - be it the ATM-based B-ISDN or whichever network else.

Leading edge science should take the approach of 'interesting vertical pilot-projects' to establish broadband networks ahead of what is available as a standard service from telecoms. The Gigabit Network Testbeds and Stuttgart VBN trials are such examples.

ACKNOWLEDGEMENT

The DFN association has supported the Stuttgart VBN and RPC projects by grants from the Ministerium für Forschung und Technologie under the titles DFN CXLX (TK 558 HD001) and DFN RPC (TK 558 VA001).

The VBN multiplexers - E-Mux and F-Mux - were developed by Prof. Kaiser, Institute for Data Transmission, University of Stuttgart.

LITERATURE

[1] ALCATEL ACC: Working Group on Connectionless Transport Service. August 22, 1990

[2] ALCATEL: Alcatel MAN The QPSX Connection. 1989

[3] Telecom Australia Research Laboratories: DQDB MANs Timetable. 1990

[4] Bellcore Technical Advisories on SMDS, TA-TSY-00072 - TA-TSY-00075, October, 1989 - March, 1990; Supplements December 1990

[5] Biersack E., Feldmeier C. (Bellcore): Transport Protocol Issues for ATM-based Networks; in [EFO90]

[6] Brownstein Ch.: US National Research and Educational Network

[7] Casner S., Seo K., Edmond C., Topolcic: N-Way Conferencing with Packet Video. USC/Information Sciences Institute, BBN. 1990

[8] CCITT: Draft Recommendation I.122: 'Framework for providing additional packet mode bearer services'. CON XVIII-R 48(C)-E, Geneva, February 1988

[9] CCITT: Recommendation G.707 - G.709

[10] CCITT: Recommendation I.122: 'Broadband Aspects of ISDN'. Blue Book, Geneva Switzerland, 1989

[11] CEC, XIII/F/GE0182: Operation 1992 Investigation of requirements and options in the field of Advanced Communications-technologies in Europe. 7 May 1990

[12] Chao H.J., Johnston C.A.: Asynchronous Transfermode Packet Video System. Optical Engineering, Vol.28(7), July 1989

[13] Christ P.: Visualization of Large Scale Simulations in a High Speed Network Environment. RARE European Usr Meeting on High Speed Networking. Brussels, February 1989

[14] Detecon Deutsche Telepost Consulting GmbH. Technisches Zentrum Beriin: BERKOM - Telecommunication within the Optical Fiber Network. Version: March 1990

[15] Domann G. et al.: B-ISDN Field Trial Concept. ISDN Europe 86. IGI Europe. Basel November 5-7, 1986

[16] Edmond W., Seo K., Leib M., Topolcic (BBN): The DARPA WidebandNetwork Protocol; in [ACM90]

[17] Fischer W., Goeldner E.-H.: The Evolution from LAN/MAN to Broadband ISDN. Draft Contribution to ICC '91, Denver 1991

[18] DEC, Northern Telecom, StrataCom(, CISCO): Frame Relay Specification with Extensions based on proposed T1S1 Standards. Document Number 001-208966, Revision 1.0, September 18, 1990

[19] Hemrick Ch., Klessig R. W., McRoberts J. M.: Switched Multi-megabit Data Service and Early Availability Via MAN Technology. IEEE Communications Magazine, April 1988 - Vol. 26, No. 4

[20] IEEE 802.6: Draft Proposed Distributed Queue Dual Bus (DQDB). Metropolitan Area Network. D6 1988 - D15 December 1990.

[21] IEEE Computer: Gigabit Network Testbeds. September 1990

[22] Kirstein P.: Upgrading the Intercontinental Links for Research with emphasis on the UK-US Scene. Computer Networks and ISDN Systems. Special Issue: EARN/RARE Joint Networking Conference. Killarney, Ireland, 15-17 May, 1990

[23] Kirstein P.: Upgrading the Intercontinental Links for Research Colloboration: Fat Pipes. iesnews, Issue No 31, December 1990

[24] Lang U., Ruehle R.: Scientific Application Environments. SIGGRAPH 90 Workshop on Data Structures and Access Software for Scientific Visualization. August 1990

[25] Lai W.S.: Frame Relaying Service: An Overview. Infocom '89, Ottawa, 24-27 April 1989

[26] Lemke H.U., Rhodes eds.: Computer Assisted Radiology. Proceedings of the 3rd International Symposium CAR '89. Berlin, June 1989

[27] Piscitello D., Lawrence J.: Internet Draft IP and ARP over SMDS. November 1990.

[28] RACE: R&D in Advanced Communications technoogies in Europe. Workplan '91. Background Material - Rationale & Overviwe, Definition of Scope and Task Descriptions. Draft Status: December 1990

[29] Ruehle R., Lang U.: Scientific Applications in a Supercomputer Environment. Keynote on the Fifth International Symposium "Science and Engineering on Cray Resaerch Supercomputers", London, October 22-24, 1990

[30] Russel T. (UltraNetwork Technologies): UltraNet High Performance TCP/IP. July 1990

[31] Siemens: Metropolitan Area Networks. Product Description. October 1990

[32] Siemens: Digitale Nachrichtenübertragung 2. Crossconnect - Multiplextechnik. 1990

[33] Sincoskie D., Bellcore: Private communication. October 1990

[34] US SPRINT: The Frame Relay Solution. Tele Notes, Volume 1 No.2, September 1990

[35] Topolcic C. (BBN): ST II. private communication December 1990

[36] Zarki M. El, Shroff N,: Performance Analysis of Pcket Loss Recovery Schemes in Innterconnected LAN-WAN-LAN Networks. Third IFIP WG 6.4 Conference on High Speed Networking. Berlin, March 18 - 22, 1991

Fast Access to Supercomputer Applications

K. P. Görtz, K. Schmidt

Ultra Network Technologies GmbH, Max-Vollmer-Str. 1, W-4010 Hilden, FRG

Abstract

Scope of this article is to discuss traditional versus state-of-the-art approaches in high speed data communication featuring the UltraNet™, connecting a variety of supercomputers, mainframes or workstations. Typical network topologies and installation profiles are shown, and one example of a distributed application in the field of chemistry is mentioned. Finally, trends in the rapidly growing network market are analyzed.

Introduction

The last decade showed a sustained high level of growth concerning all parameters of computer hardware. CPUs of entry model workstations perform 10 MIPS, physical memory for near supercomputers is in the range of 2 Gigabytes, clockrate, degree of parallelism are only mentioned. With sophisticated techniques like caches or interleaved memory, the high performance CPUs or vector units have to be loaded with data and the computed results are to be stored or to be visualized without occurring bottleneck.

Unchanged in the last decade is the users impatience and his wish for minimal system reaction time regardless of whether his keystroke displays a character string or triggers the backup of a 100 MByte file.
For a remarkably long time Ethernet's 10 MBit per second was taken as the *upper networking limit* and 10% or 1 MBit/s thereof, when measured effectively from userbuffer to userbuffer, was common. The network as the weak point of the whole configuration was widely accepted. Using data compression algorithms or data reduction in the supercomputer application to fit the weak network throughput, one tried to prevent the further opening of the "network scissor" between Inside and Outside. Just recently high speed products like protocols, channels and hereto tailored applications are proposed and furnish the basis of ongoing development. A major point is the presence of powerful host channels with speeds up to 800 MBit/s as there are networking technologies like FDDIs 100 MBit/s or the UltraNets GigaBit bandwidth (see Fig. 1).
The traditional usage of a centrally operated computer center service is changing remarkably. Among others the usage of desktop-computers like PCs or workstations offer the whole power of the computer center at the users desk. Both computers of different vendors and communicating scientists or institutes share the common resources. Centralized or decentralized file servers, visualization devices, or special purpose computers are networked and are at the remote users disposal. Increasingly, distributed applications find their usage,

U. Harms (Ed.)
Supercomputer and Chemistry 2
© Springer-Verlag Berlin Heidelberg 1991

which, with client and server methods (i.e. via remote procedure calls) separate the compute intensive task on the supercomputer versus the graphical tasks on the workstation. Todays databases or programs like NFS support distributed data.

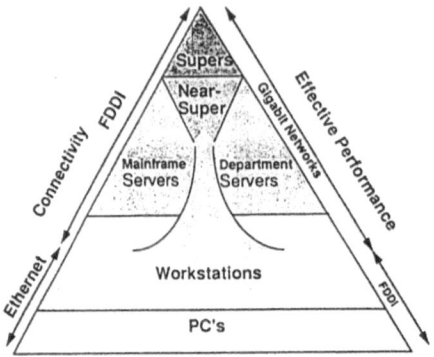

Fig. 1 - Technical Network Market.

Figure 1 shows the typical existing pyramid of networked hosts, ranging from desktop PCs via departmental computers -as workstations or nearsupers including local subnets - up to centrally operated computer centers with supercomputers, mass store devices or video/print output devices.

The wish of the user, unwilling to learn languages, operating system or networking tools leads unavoidably to the usage of standards. Operating systems like Unix - applicable from PC to Supercomputer- facilitate the transparent access towards data or system functions of remote hosts. Program compatibility at least on the graphical user interface level is required. Data security and virtual access to huge amounts of data is a service the user will be asking of his computer center. Especially when integrating super computers into the network, pixel oriented framebuffers or object oriented workstations are used to interactively manage the run of the super computer or to offer the enormous amount of output data in a human understandable form. Animated high quality graphics is used more and more in diverse fields to visualize complex phenomena (see Fig. 2).

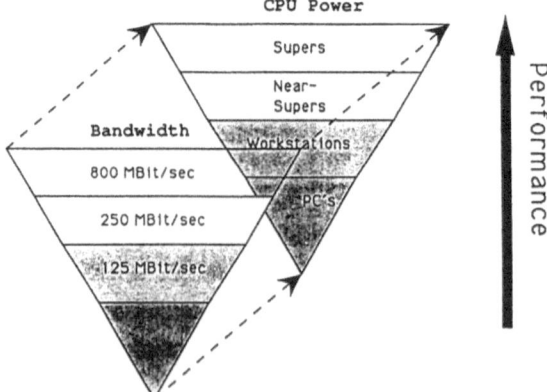

Fig. 2 - I/O Bandwidth matches CPU power.

To match the diverse requirements of network throughput (i.e. text 9.6 kbit/s ; animation 1000 Mbit/s). Fig. 2 shows in the network pyramid the match of compute- and network power.

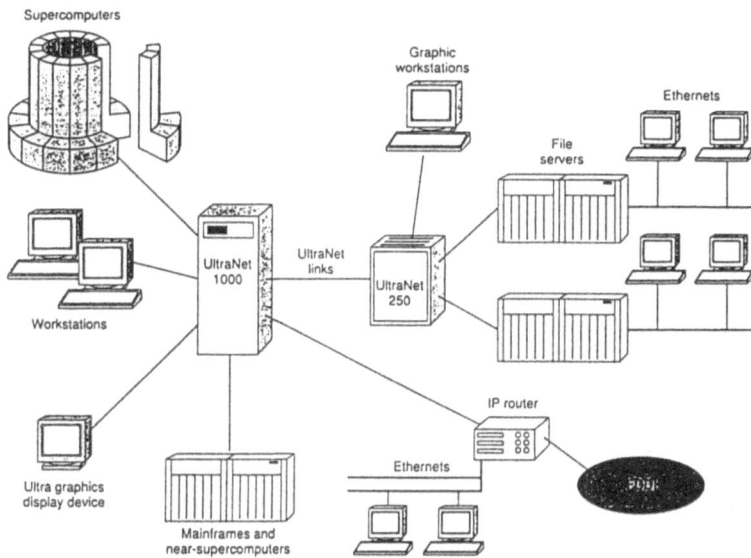

Fig. 3 - UltraNet Topology.

UltraNet is a HUB oriented packet switching LAN with a bandwidth of 1 Gigabit/s per hub. The 64 bit wide computer-like UltraBus runs internally as a conflict free broadcast LAN and controls the transfer between adapters:

Host adapters connect the UltraBus with the I/O-channels of the host (i.e. Cray HSX or IBM HIPPI) furnishing connection oriented and datagram services following the OSI layers 1-4 in hardware.

Link Adapters route packets as MAC bridges from one hub to the other, to gateways of other networks or to the target machine via serial coax or fiber optic lines. The 4B/5B encoding scheme used the same in FDDI is used alike within the UltraNet with 62,5 125 and 250 MBit/s link speeds. Four physical lines can be stripped to one logical to permit a maximal link bandwidth of 4x250 Mbit/s.

A Framebuffer presents the pixel oriented content of the hosts memory in realtime on a color screen. The framebuffer is an output device only and is addressed as a normal UltraNet host.

Hub resident adapters use the fastest I/O channel of their host for channels as the HSX and HIPPI with 800 MBit/s and the BMC with 24 or 36 MBit/s and are located from 25m to 120m away of the Hub (without channel extensions). Host resident adapters like the VME adapter connected to the Hub via serial lines (coax 75Ω; fiber 50μl/62,5μ or 8μ monomode) and can be as far as 30 km away without repeater. Figure 4 gives an example of a typical UltraNet installation profile including connection to medium or lower speed networks like Ethernet, Token ring or FDDI.

Central points within the UltraNet are the Hubs UN1000 and UN250 (VME machines only), hosting all adapters except the host resident adapters. Redundant inter hub links are possible.

The general Purpose Router AGS+ from cisco systems with the jointly developed Ultra/cisco interface realizes the attachment of subnets like several Ethernets, DecNets, Token Rings or FDDIs. Accordingly UltraNet is able to serve as a backbone to these subnets. UltraNet enables (due to the external protocol processing) effective rates from 30 to 50 %, even 90% in conjunction with the Ultra Framebuffer, of the nominal channel bandwidth. Another important feature of the external processing of the protocol is to offload the host CPU. Users can't afford to misuse expensive Cray, Fujitsu or IBM CPUs for ineffective generation of network packets. UltraNet combines the use of backend and frontend functions and can thus be used as a universal network or as a pure backend for FDDIs or Ethernets.

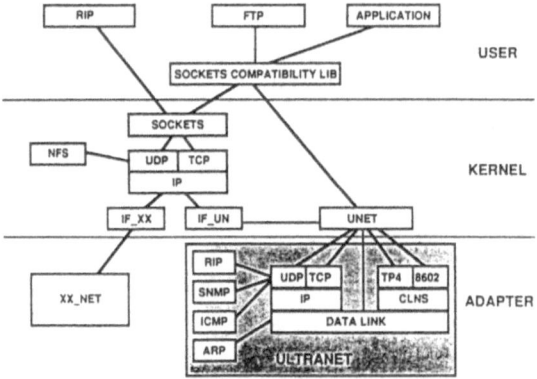

Fig. 4 - UltraNet Dual Protocol Software.

UltraNet host software is based on five development principles (see Fig. 4).

1) Efficient coupling of UltraNet adapter and host.
2) Efficient support of a flow-through-architecture, elimination the classical model of store and forward.

3) Portability onto a variety of platforms and operating systems.
4) Maximum support and integration of industry standards like ISO TP4 and TCP.
5) Cooperation with existing - host resident- network services.

The adapter driver resides in the kernel of the operating system. By ignoring the host resident networking features, internal bottlenecks like copying in the kernel space, or kernel management can be bypassed. To assure the portability of the kernel software from scratch the software is divided into a logical (system independent) and a physical (system dependant) module. Both components are standard Unix drivers, communicating directly with the user processes via system calls. Small request blocks - carrying no user data - deliver messages between the adapter and the user evoked process. The actual data packets (TPDUs in OSI terminology) are up to 32 Kbytes in size and are not buffered on any adapter, but run directly from source to sink userbuffer via the link or UltraBus.

A linkable socket compatibility library (SCL) on OSIs session layer multiplexes the use of UltraNet or Ethernet (or any other host processed network). The SCL is the actual programming interface for existing or user programs. Standard network applications like FTP are only to be relinked with this library.

In case the target host is known (using standard internet style addressing) as a native UltraNet device (i.e. a framebuffer) the UltraNet driver unet is called directly. Otherwise the kernel implemented BSD socket interface is used (not the Ultra-socket interface) which returns the desired file descriptor. Programes like NFS or host internal mail services which cannot be accessed via the socket interface, are given a second way to communicate over the UltraNet. IP packets are given to the Ultra driver via another (Ultra over IP) interface, encapsulated from the network processor into OSI packets, routed through the UltraNet towards the target adapter. In the receiving host the striped IP packet is further processed by the internal IP protocol.

Recent implementations of PP and adapter software, contain, apart of Ultras ISO implementation, the protocols of the TCP/IP family, both hardware-supported. The use of the protocol is dependent on the target machines address. Between hubs, UltraNet hosts and framebuffers the more powerful OSI protocols will still be used. When explicitly desired or when communication with non Ultra host (via the IP router) TCP will be processed.

As already mentioned, the interface operating system-network is of significant importance. Needed for a balanced overall performance, this interface is hardly considered by traditional networks, only offering data link services to the host. With other words:" To bring and get data from and to the network is part of the responsibility of the host"!

UltraNet offers hardware supported transport layer services to the user or the user process. With this approach high data rates of about 50 Mbytes/s between user application buffers of two HIPPI-hosts are achievable. A high bandwidth at the link level, basically profitable for simultaneous sessions, is only one part of the story.

Installation profiles

Following are three examples of UltraNet installations profiles in the domain higher education and research :

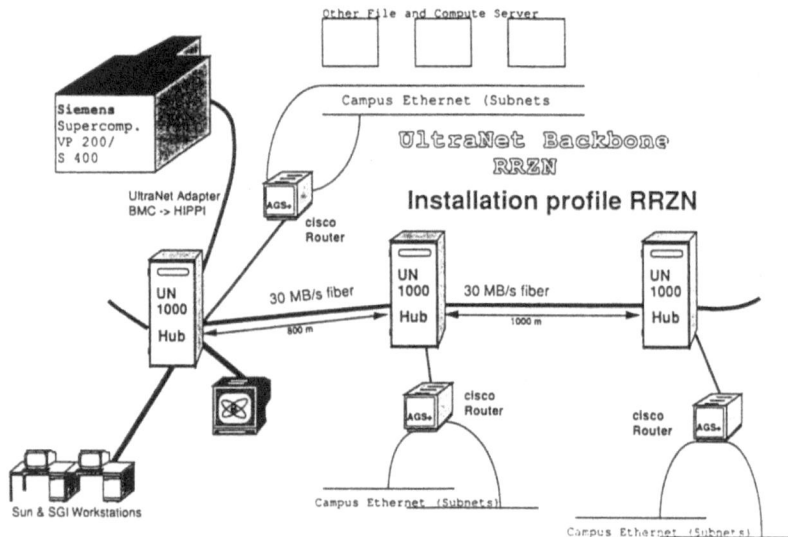

Fig. 5 - RRZN Profile.

Figure 5 shows the UltraNet installation of the regional computer center from university of Hanover, just about to be set up. The main computer is the SIEMENS vector model VP200-EX, soon to be replaced by the S400. Initially BMC channels are being used until the implementation of the HIPPI channel is finished. This port is being done by SIEMENS in München. Connected to the super computer are a framebuffer, dedicated workstations as well as Ethernet segments via the cisco router. FDDI is not foreseen. The aim of RRZN is the implementation of a universal network in the sense of OSI, which supports high speed demands.

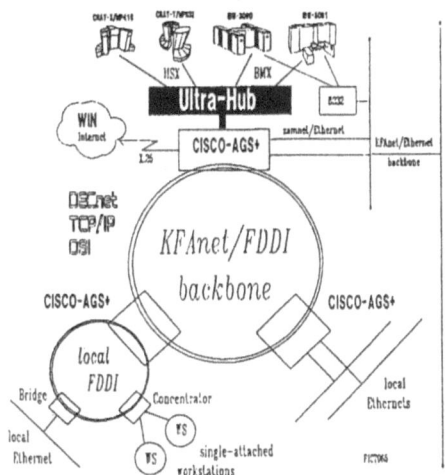

Fig. 6 - KFA Profile.

The second example shows the network concept of the research center in Jülich (see Fig. 6). At present 2 large Cray systems with 100 MByte/s channels are connected via the UltraNet. The next planned step is to add 2 IBM mainframes. Cisco routers are used to serve medium and lower speed networks. The backbone here is an FDDI-ring whereas UltraNet is used as a high speed backend network. This concept shows the useful integration and enhancements of UltraNet and FDDI.

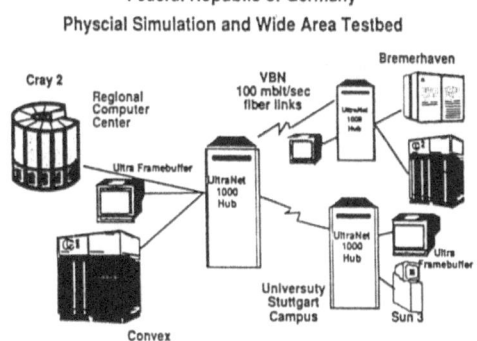

Fig. 7 - RUS/AWI Profile.

Figure 7 illustrates the first European UltraNet installation at the University of Stuttgart and the Alfred Wegener Institute for Polar and Marine Research in Bremerhaven. Both local UltraNets are connected (within the framework of a German PTT VBN pilot project) with 100 Mbit/s. The Stuttgart based Cray 2 is used as the compute server for the Bremerhaven Alliant machine and the Ultra Framebuffer (see Fig. 8).

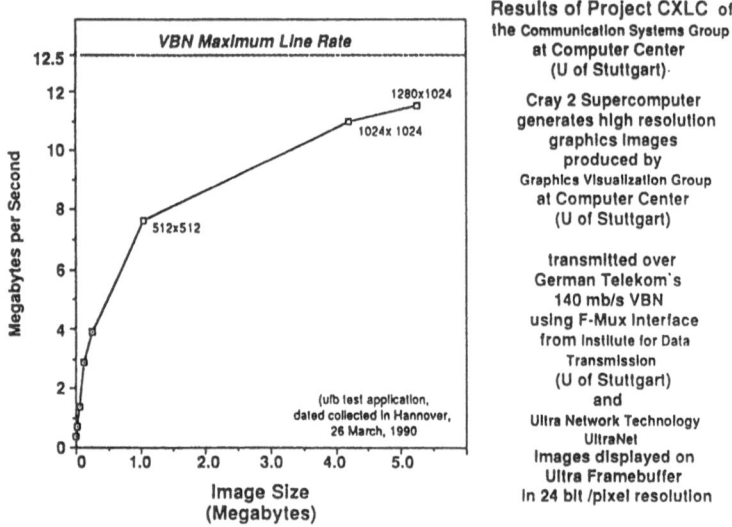

Results of Project CXLC of the Communication Systems Group at Computer Center (U of Stuttgart)·

Cray 2 Supercomputer generates high resolution graphics images produced by Graphics Visualization Group at Computer Center (U of Stuttgart)

transmitted over German Telekom's 140 mb/s VBN using F-Mux Interface from Institute for Data Transmission (U of Stuttgart) and Ultra Network Technology UltraNet Images displayed on Ultra Framebuffer in 24 bit /pixel resolution

Fig. 8 - VBN performance: RUS and CeBIT Telekom Booth.

During the CeBit fair in Hanover 90 Mbit/s could be seen for image display. The access of the supercomputer disk 600 km distant was much faster than even from the local SUN workstation.

This common project of University of Stuttgart, the German PTT and Ultra has received worldwide and considerable interest. Within this project a number of performance measurements have been done (see Fig. 9).

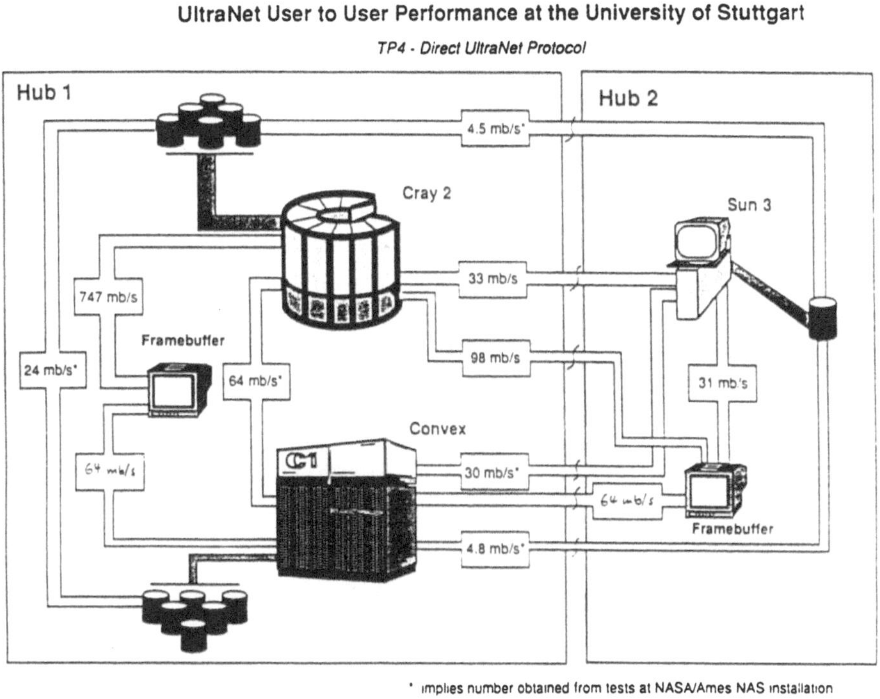

UltraNet User to User Performance at the University of Stuttgart

TP4 - Direct UltraNet Protocol

* implies number obtained from tests at NASA/Ames NAS installation

Fig. 9 - RUS performance measurements.

Most supercomputer applications of today are typically run in batch mode following the philosophy: "The computer should calculate and not be bothered by users". This philosophy leads to way things are: a preprocessed input file is given to the supercomputer and the algorithm is processed without user interaction during a notable amount of time. Spot checks regarding numerical instabilities and input errors or the possibility of interactive change of parameters are not possible. Final results are stored on disks as huge output files, if the data reduction is not done in the same computer run. The interpretation of this stored data (post processing) is most often done on interactive graphic workstations. Usually the results are too large to reside in the local workstations. With transparent access to super computer disks or local server via slow Ethernet type networks with file sharing programs like NFS, huge data interpretation is at least troublesome if not impossible. This leads to the fact, that the scientist, anticipating results, looks for them exclusively trying to reduce the volume of the data. In batch mode subtle effects, time or other dependances are difficult to detect within the run. The possibility of scanning raw unfiltered data in real time, by different graphical interpretation algorithms interactively controlled by the scientist, delivers a new way of understanding and increases the value and quality of super computer runs.

Therefore, effective throughput in the 90 Mbyte/s range is desirable to guarantee a high quality animation with 1280 x 1024 bits with 24 Bits color information (4 Mbytes), and up to 20-30 frames per second. Using the UltraNet Framebuffer this type of application is possible (see Fig. 10).

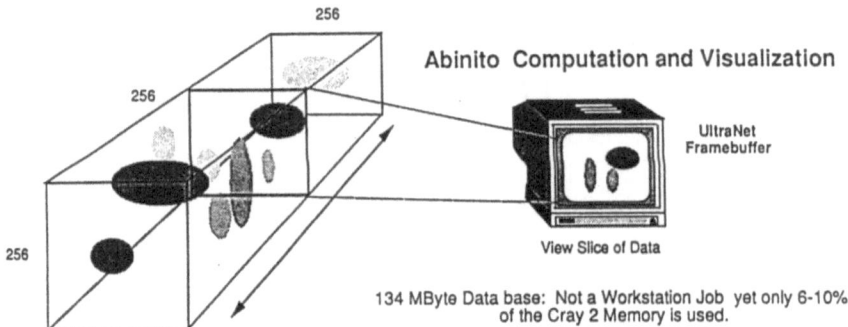

256
256
256

Abinito Computation and Visualization

UltraNet
Framebuffer

View Slice of Data

134 MByte Data base: Not a Workstation Job yet only 6-10%
of the Cray 2 Memory is used.

Examine very large data base for quick understanding of results

1. 2D density structure at any Z point (color specifies density range)
2. Change density range to see details.
3. View 3D by applying "depth of field" colors

Fig. 10 - "Walk through data". Visualization Example.

As an example of interactive computing a 3 dimensional data set with 256^3 gridpoints containing double precision floating point values, are analyzed.
The ab-initio output of electron density of a molecule represents a medium resolution problem with 128 MBytes/ 60000 printed pages. Even one slice of this data cube contains 0,5 MByte, which still leads to a respond time of several seconds, using data access via traditional networks.

A walk through the density field, color contrast amplification, magnifying or hiding of discreet regions, 3 dimensional representation from every perspective or the usage of mathematical filters, require a continuous data flow to depict dependances.
In conjunction with a scan converter interesting image sequences can be brought to video. So the results can be published and made accessible to the public.
More elegant would be the following: Most super computer applications store their interim results of gridpoints in the super computer memory. One additional process, knowing address and structure of this memory location, reads this data asynchronously in arbitrary intervals and displays it on the networked monitor. The simplest form of controlling the programs behavior, is to stop when necessary. Input errors in the configuration files, numeric or algorithmic inconsistencies are visible immediately and not after hours of having the

supercomputer calculated 100 Mbyte of zeros. Under certain conditions the application is able to accept interactive input to control the program run. The premature recognition of stable results can thus save a good amount of expensive compute time.

Increasingly, concepts like the client server model find their use. By direct process to process communications for instance with remote procedure calls supercomputer applications are separated into compute and graphic intensive parts. With effective data rates from 3 to 13 Mbytes/s a 2D workstation can be delivered with pixel, a 3D workstation can be delivered with graphic objects with calls to a distributed graphic library. This range of network performance allows several pictures a second, if size or color depth is scaled accordingly.

The final goal of these effects are to give the user a tool to simulate reality, and to see the reactions of his system's parameters to interactive changes, directly and comprehensively.

Perspectives

Looking back in time, access to supercomputer applications was realized via frontend and mainframe systems in the 70's. In the mid 80's direct access from workstations to supercomputer became reality. In the 90's user data and compute resources will be distributed throughout the network between conventional super computers, massive parallel systems, networked file servers or workstations (the user's interface into the network).

Two important key elements to this are the implementations of fast file servers as well as data-management-systems.

File servers act as buffers between networks or hosts. They store data for later postprocessing or visualization (possibly on video tapes). In the near future effective transfer rates of 20-40 Mbytes from and to the file server will be possible. But in the domain of visualization demands of 100 Mbytes/s occur.

With modern data storage management systems (i.e. Unitree or data tree) one significant disadvantage of desk computers is at least reduced : small desk storage. A variety of high end workstations or mini super computers are capable, if connected with high performance networks, to run memory demanding applications with very fast access to network file server, which could not perform satisfactorily with local disks. High speed networks allow the staff of a computer center to run more cost effective special purpose devices. The graphic workstation must not have super computer power, the supercomputer must not have it's own high performance disks. Powerful visualization-, vector-, file-server or parallel systems are at the users disposal. The balanced supercomputer on every users desk is then a reality.

High Speed Networking Solutions

J. W. Adams, G. Hemminger
Network Systems GmbH, Lyonerstr. 30, W-6000 Frankfurt 71, FRG

Abstract: Today, open networking systems based on standards are reshaping the computer and networking industries. This paper discusses some of the ways organizational computing and networking have changed in recent years, what customers are looking for, and how companies like Network Systems are changing and creating a new generation of networking to meet those needs.

THE CHANGE IN SYSTEMS

Figure 1 gives an overview of where Network Systems saw the EDP industry in the past. Large-sale computing was characterized by hierarchical, closed system architectures, proprietary interfaces, and limited multi-vendor connectivity. There were few networking vendors, and products were expensive. Data processing operations themselves were centralized, involved a relatively small number of people, and served primarily in a support role. As CPUs became more powerful, input-output and networking became the bottleneck.

- Closed System Architectures

- Proprietary Interfaces

- Limited Heterogeneous Connectivity

- Expensive Products

- Few Competitors

- Major Growth in Mainframes/Minicomputers

- Mini Supercomputers are introduced

- Supercomputers become commonplace

- Networking becomes the Bottleneck

*"A Compute-Bound Problem Becomes an
I/O-Bound Problem"*

Fig. 1 High Performance Networking - The Past

U. Harms (Ed.)
Supercomputer and Chemistry 2
© Springer-Verlag Berlin Heidelberg 1991

Figure 2 recaps how we see present trends developing at our current customers. The situation is vastly different. Open systems architectures and standards-based technologies are the preferred choice for many users. Many more vendors are supplying multi-vendor connectivity of various kinds, and the price performance of these products has improved greatly. Distributed computing has increased with the explosive growth in PCs and workstations, while mainframe business has slowed and the role of the large CPU has changed. In supercomputing, the demand for number-crunching power keeps increasing, along with the domain of solvable problems. But high-speed networking standards are helping to change the way users work with supercomputers, and are turning previously I/O-bound problems into computer-bound problems once again.

- Open-System Architectures
- Standard-Based Technology
- Improved Price-Performance
- More Competition
- Rapid Growth in Workstations
- Mini Supercomputer become Application Servers
- Supercomputer demands Expand
- Growth in Mainframe Flattens
- The System Solution is Networking

*"A Compute-Bound Problem will again become
a Compute-Bound Problem"*

Fig. 2 High Performance Networking - Present Trends

WHY NETWORKING?

Today, in fact, the systems solution is a networking solution. Networking is the glue that holds distributed systems together. Organizations are finding their networks to be critically important, both as a way of saving money through sharing resources, and as a source of competitive advantage. They can get the needed information to the right people in a timely way, for every sort of purpose from better decisions on pricing or inventories, to more computer simulations of a new product such as a drug compound. This new world of open networking systems has brought with it a new set of issues and its own complexity. As networks have become more critical to organizations, they are applying a more rigorous set of criteria. These are connectivity, performance, support and service, ease of installation and use, standards-based product, and comprehensive solutions. Buyers also look at the experience and financial stability of the vendors very carefully. These criteria add up to a requirement for a NEW GENERATION OF COMPUTER NETWORKING.

HIPPI

We think the elements of a successful strategy for dealing with this new generation include providing continuing increase in performance, supporting a core set of industry standards, offering a number of comprehensive solutions, not just boxes, and taking a systems approach to customer requirements. By higher performance we mean 100 megabit per second, 800 hundred megabit, and gigabit networking products. Network Systems, for instance, announced its P8 HIPPI Switch, as depicted in figure 3, earlier this year.

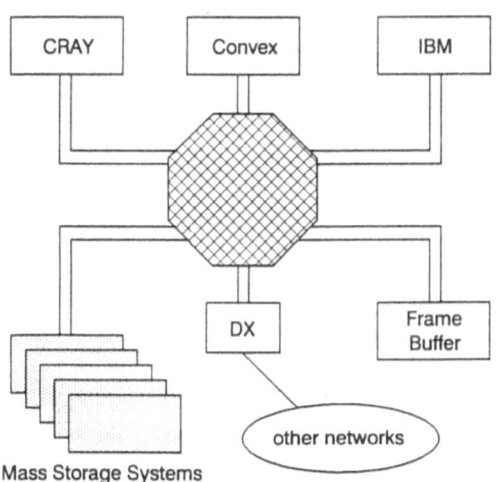

Fig. 3 P8 HIPPI Switch

The P8 is a crosspoint switch with a total throughput capacity of 6.4 gigabits which allows 8 simultaneous dual-simplex conections as shown in figure 4. Just recently we introduced a follow-on product, the PS32 Switch, with up to 32 source and destination ports and an aggregate capacity of 25.6 gigabits per second. The PS32 is the result of a collaboration with Cray Research.

- up to 8 simultaneous Dual-Simplex Connections
- full capacity Capability of 6,4 Gbits/sec

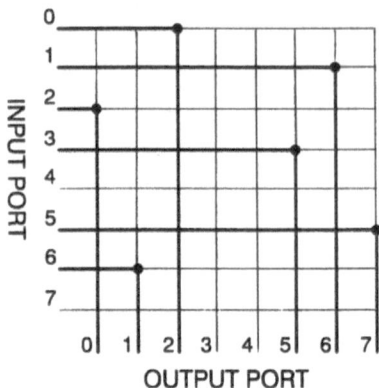

Fig. 4 P8 HIPPI Switch Connections

Network Systems and Cray jointly developed the specifications for the switch; Network Systems built it; and Cray is providing complementary software for it. Cray will also provide complementary software for the new version of the earlier switch, the PS 8-8.

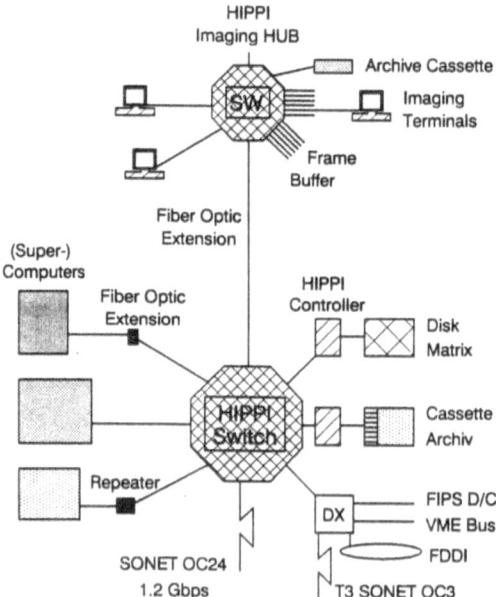

Fig. 5 HIPPI Networking Directions

This kind of alliance will become increasingly prevalent, we think, as vendors cooperate to provide effective solutions for customers, allowing even greater efficiencies for higher speed cpu's, PTT links, peripherals and subnets as in figure 5. High performance also means better network availability, enhanced network management, and having the network handle additional functions such as protocol processing, to free up CPU cycles.

STANDARDS

When it comes to standards, the old joke was that they must be good, because there are so many of them. Figure 6 gives a few examples.

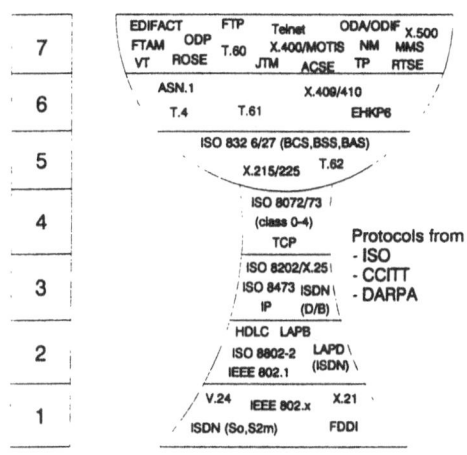

Source: Computerwoche

Fig. 6 OSI - Which Standard Do You Want?

By one count we now have some 250 committees working on over 1,000 standards of various types. In our experience, networking vendors have to focus on a core set in order to function effectively. Network Systems has elected to focus on FDDI, Ethernet, TCP/IP; OSI, DECnet, HIPPI, and Token Ring. This allows a geographically hierarchical network structure, based on standards, as in figure 7. To further this approach, Network Systems recently introduced two new router models based on the FDDI standard. As shown in figure 8, one connects two FDDI rings, and the other connects an FDDI ring to a T3 communication line. At the same time, just complying with standards is not the complete answer. We know that from our participation in several tests and trade shows, that committees sometimes move at a glacial pace, and that achieving interoperability based on standards can often take a fair amount of effort. Of course, there is always the pressure to meet a need today rather than tomorrow.

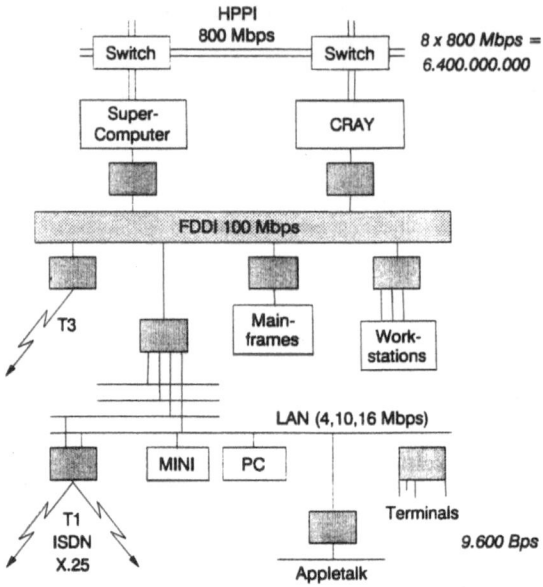

Fig. 7 High Speed Networking in the 1990's

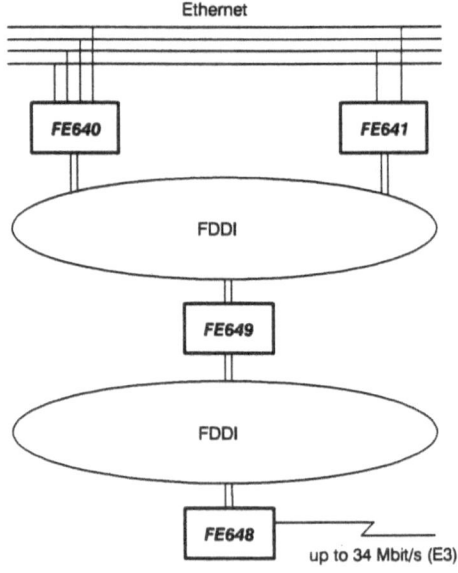

Fig. 8 FE6xx Router

QUICK RESPONSE TO SOLUTION NEEDS

Vendors therefore need to do more to provide quick comprehensive solutions, based
on these standards. We began with the development of our Data-Exchange (DX)
architecture, which has allowed us to attach new standard interfaces quickly.
With this structure, figure 9, we were able to develop single and dual attached
FDDI board sets, as an example, and provide immediate solutions such as the
simple FDDI-FDDI-Host connection in figure 10 or the complicated bridging and
routing of the networks and connections of hosts as in figures 11 and 12.

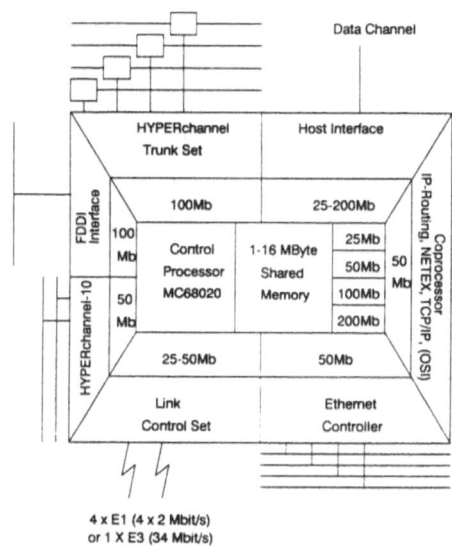

Fig. 9 HYPERchannel DX Architecture

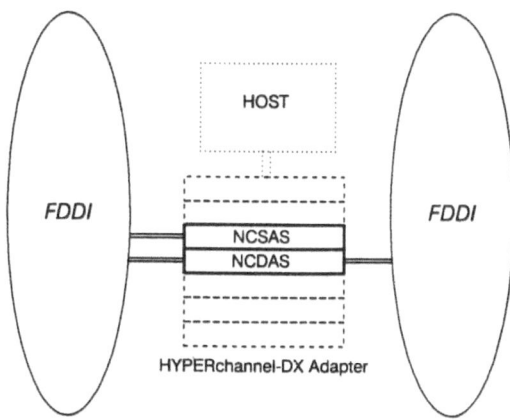

Fig. 10 HYPERchannel-DX FDDI Interface

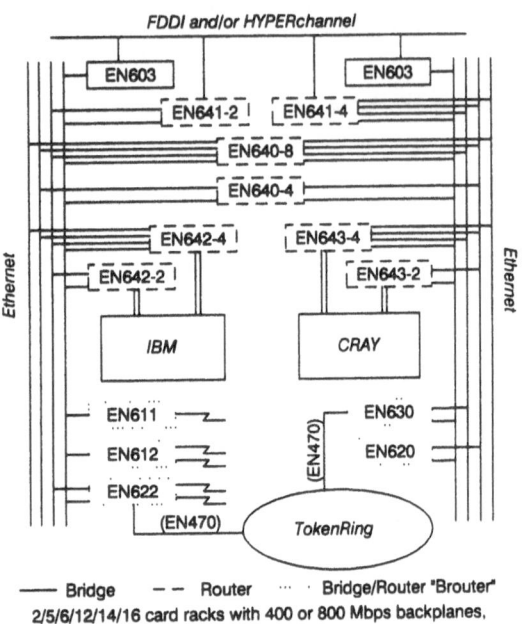

Fig. 11 EN6xx Bridge/Router In Production

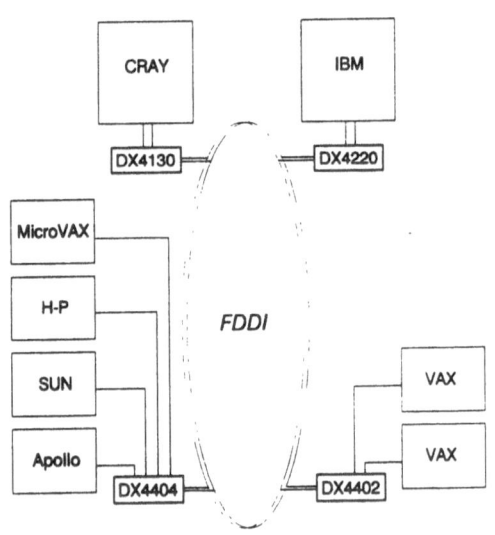

Fig. 12 DX-4000 Series (dual attached)

Network Systems also offers application software for functions like centrally attended backup of remote systems, automated file transfer, a wide range of

connectivity for hosts and peripherals from some 40 systems vendors. We have
also developed routers of industry leading speed, handling a wide range of
protocols and media as shown in figures 13 and 14. We provide worldwide support
and service, including 24 hours a day, on demand.

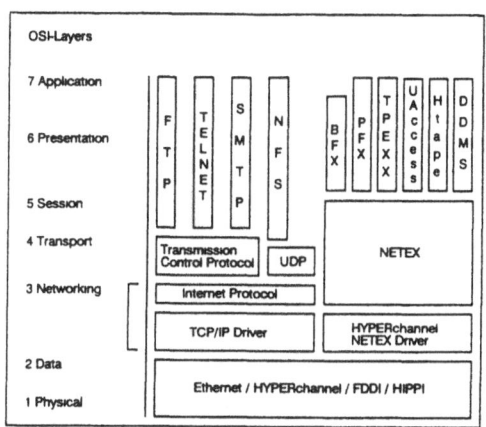

Fig. 13 Network Systems-Software and the OSI-Reference Model

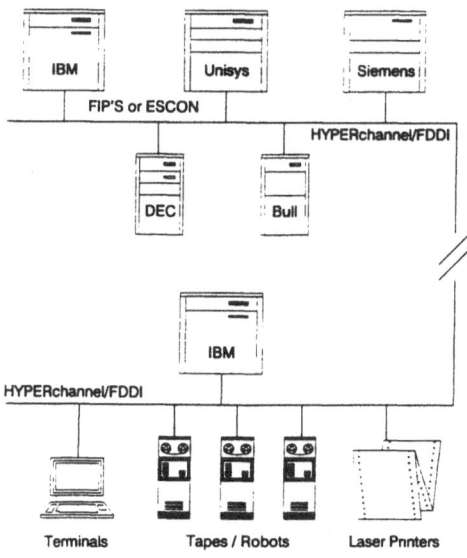

Fig. 14 HYPERchannel connects Computers and Computer Centers

THE NETWORK SYSTEMS PRODUCT SEGMENTS

To do a better job of providing these solutions, Network Systems focuses on three
major areas both in development and marketing: technical computing,
internetworking, and commercial computing. Here is a brief look at some of the
products in each area.

Supercomputing

In the supercomuter segment of the technical computing market, the HIPPI standard
is the basis of our new products. The P8 and P32 HIPPI Switches, the
highest-speed commercially available devices for computer networking, can be used
to connect supercomputers with each other, with large data storage devices, and
directly with privileged workstation users. The switches are complimented with a
full range of of HIPPI connection and application products to insure you can use
them today. The INI (Intelligent Network Interface) allows permanent long
distance connection between HIPPI switches so that there is minimal connection
latency. This allows full motion graphics without sitting in the computer
center. Our HIPPI chipset, produced jointly with AMCC, is allowing vendors such
as Sun to quickly provide full HIPPI standard high speed graphic connections. Of
course, our chips are also used in a connection to DX so that we are downward
compatible to our current software and other hardware lines. We also have SONET,
Fiber Channel, and VME connections in various states of test and delivery. For
these reasons, we were recently chosen as the networking partner by DARPA and
Carnegie-Mellon for the NECTAR supercomputing project. This project advances
supercomputing to a point where the network itself is the supercomputer,
connecting many application specific parallel processors totally transparent to
the user.

Internetworking

For the Internetworking segment of our business, we introduced a series of
routers and bridges based on the FDDI, ethernet and token ring standards to
support distributed computing. One of the consequences of distributed computing,
however, is that many important files are now created away from the central data
center, on workstations or minicomputers. But out in these departments there may
not be the people, peripherals, and procedures to ensure the routine backing up
those files, in case the original is lost. To meet that very real need, Network
Systems introduced Central Archiving, which enables an organization to
automatically back up data from dissimilar systems at remote sites into a
controlled central evironment. Central Archiving is a feature of both HYPERtape
and USER-Access, software that works with NETEX (Network Executive) communication

protocols. NETEX is an early implementation of the OSI protocols and therefore supports heterogeneous networking. Central Archiving can save personnel costs and hardware costs while it speeds up processing and improves backup management. Users of DECnet or TCP/IP networks can now take advantage of USER-Access, for interactive and batch file transfers, for executing commands on remote systems, and for centralized file backup.

Commercial Computing

In the world of commercial computing, which is dominated by IBM, we offer a number of complimentary products. In the channel extension and channel networking market, we supply a range of products from the low-priced 9300 fiber extenders for short-distance relocation of peripherals, to the 9520 Enterprise Channel Extender for the urban market, to the 9700 Enterprise Channel Controllers for complex, long- distance application. This broad range of products allows companies to run the same applications they used in the past, but over extended distances without performances loss. Widely seperated equipment can be treated as local. In additon, the 9700 family allows a company to add multi-vendor networking capabilities as needed, which can mean the introduction of new application. In addition, for the commercial user, Network Systems announced a new product for automated file transfer and management. DDMS, or Data Delivery Management System, allows companies to automate the scheduling of transfers of data, get round-the-clock use of existing mainframes and data centers, and do this over multiple networks, including SNA, HYPERchannel, and X.25. DDMS is an implementation of an open systems architecture for use in homogeneous environments.

THE CHALLENGE

In short, the challenge for companies like Network Systems is to adapt to the new requirements of a new generation of networking, by focusing on standards-based products, higher performance, greater interoperability, and comprehensive solutions. While no one rings a bell to announce these new eras, these changes have certainly been quite evident now for a couple of years. It is reasonable to infer from Network Systems' financial results over that period that we are successfully adapting to this new environment. Our revenues have continued to climb through this period, as figure 15 indicates, in a fairly consistant pattern. Our earnings over the same period reflect a downturn in 1988 as we introduced our new standards-oriented DX technology, which is often the case with the transition from one product line to another. But the earnings since then have rebounded nicely, indicating a growing customer acceptance of this new strategy for a new generation of networking.

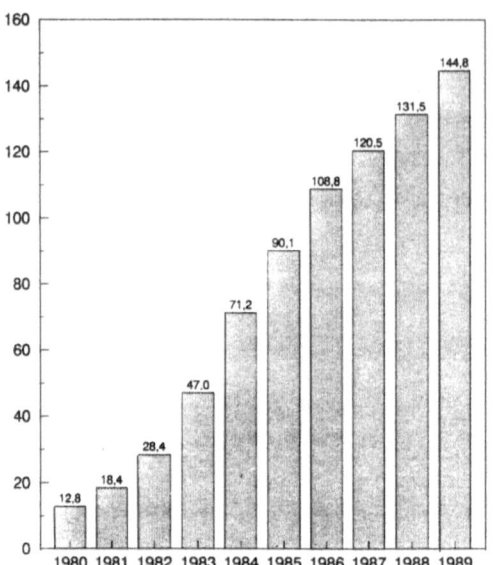

Fig. 15 Revenue Development from 1980 to 1989

Figures 16 and 17 show Network Systems is a competent networking partner for you. We provide quality solutions and service to meet your connectivity needs, whereever and whenever you need us.

- over 1100 Employees

- 11 international Subsidiaries

- 10 international Distributors

- 78 Sales and Service Offices

- over 1600 Installations

- working as primary contractor with over 70
  Software development firms

Fig. 16 Network Systems Corporation

- Performance

- Broad Product Solutions

- Experience

- Software Soultions

- Creativity

- Full Service and Support

- Financial Stability

Fig. 17 Network Systems Advantages

References

1. Altman L, (1990) Presentation for Research Consortium Conference, NSC, Mineapolis, Minnesota
2. Hemminger G, (1990) Response to CMU RFP, Nectar, NSC, Minneapolis, Minnesota

Computational Chemistry in Industry:
A Parallel Direct SCF

Stefan Brode

BASF AG, Zentrale Informatik Technologie, Kaiser Wilhelm Straße 52,
W-6700 Ludwigshafen/Rhein, FRG

Abstract: Recent sucessful applications of quantum mechanical SCF-MO calculations at BASF AG show the increasing importance of this method for industrial research and development. However, there is a need for improved performance for *ab initio* SCF calculations. The concepts for a parallel implementation of an *ab initio* SCF-program are discussed and the expected performance for a cluster of workstations is estimated.

1. Organization of Computational Chemistry Research at BASF AG

The application of computational chemistry methods to research and development has quite a long history at BASF AG. Researchers in the Farbenlabor (dye development) have successfully applied Hückel-M.O. and related methods to potential dye-compounds for over two decades. About 5 years ago drug design chemists began to use molecular modelling techniques and (semiempirical) quantum mechanical methods to explore the properties of potential drugs and their binding to receptors. Over the last two years polymer modelling and new materials design emerged as growing fields for molecular modelling and computational chemistry methods.

The research in these areas is organized in a decentralized manner. There is one research chemist in the Farbenlabor performing PPP/HMO-calculations on dyes using a DEC VAX 8820 located in the central computing center. In the future he will use a fast, local workstation (IBM 6000/320) to perform these calculations and to exploit the capabilities of semiempirical (NDO-type) and *ab-initio* quantum mechanical methods to dye research problems.

U. Harms (Ed.)
Supercomputer and Chemistry 2
© Springer-Verlag Berlin Heidelberg 1991

In the Hauptlabor a group of about 10 people use a Convex C210 minisupercomputer, a DEC Vax 11/785 with several Evans & Sutherland graphics terminals and a Silicon Graphics Personal IRIS 4D/20G to run the molecular modelling software of TRIPOS (SYBYL) and BIOSYM (INSIGHT, DISCOVER). Quantum mechanical calculations are performed with the VAMP program (T. Clark, University of Erlangen) and the TURBOMOL package (R. Ahlrichs, University of Karlsruhe). In the near future the old-fashioned Vax/E&S equipment will be replaced by several modern graphics workstations such as Personal Iris 4D/35GT and ESV 20.

The group in the Kunststofflabor (plastics department, 3 people) is using an SGI 4D/220GTXB and an SGI 4D/35TG to investigate polymers and the electronic properties of monomers and oligomers and their effect on reactions. Polymer modelling is performed with the help of BIOSYM's polymer consortium software which is currently under development. For the electronic structure calculations SYBYL is used as a graphical front end to the quantum mechanics programs MOPAC and TURBOMOL. Very recently two projects were set up in the Kunststofflabor and the Ammonlabor (inorganics department) to support x-ray structure determination with computational modelling methods. These two groups will use the CERIUS software from Cambridge Molecular Design running on two Personal IRIS 4D/35TG systems.

At the Zentrale Informatik Technologie (Central Informatics Department) a group of 4 people is responsible for the central support of the departmental modelling groups (in particular quantum mechanical methods and statistical treatment of polymers). The main goals are the application of computational chemistry methods to new areas such as zeolites, emulsifiers or diesel fuel additives and the development of computational tools which provide functionality currently not available from commercial vendors. Two SGI Personal IRIS 4D/25TG systems are used to run the software packages QUANTA/CHARMM (POLYGEN) and INSIGHT/DISCOVER (BIOSYM). For the computationally intensive quantum mechanical programs such as SCAMP (scalar VAMP), MNDO90 (W. Thiel, Wuppertal) and TURBOMOL a cluster of 8 IBM 6000/320 workstations will be used in the near future.

2. Recent applications of *ab-initio* SCF calculations

In general, quantum mechanical calculations are performed on relatively small molecular systems (less than 100 atoms) to obtain information about geometries, energy differences, electronic properties such as dipole moments or charge distributions, bond strengths, rotational barriers and reaction pathways. Most of the problems are investigated with the help of semi-empirical M.O.-methods. *Ab-initio* SCF calculations are performed on small model compounds (less than about 30 atoms) using sufficiently accurate basis sets (normally double zeta including polarization functions, dzp) in order to check the validity of the semiempirical approximations or to get more accurate information regrading energy differences. Three examples (out of about 20 projects in the last two years) will show how *ab-initio* quantum mechanical calculations may help in industry research.

2.1 Dipeptide Conformers

In collaboration with H.-J. Böhm from the drug design group we have investigated the geometries and energies of three N-methyl-N'-acetyl glycineamide (diglycine) and five N-methyl-N'-acetyl alanineamide (dialanine) low energy conformers with the help of *ab-initio* SCF and MP2 calculations using fairly large basis-sets such as dzp, tzp or tz2p. Complete optimization of geometries was carried out on SCF-level (dzp-basis) for all conformers using the TURBOMOLE program system. Our results show significant shortcommings of the standard force fields (such as AMBER, CVFF or ECEPP) normally used for protein modelling. For diglycine the *ab-initio* calculations (Tab. 1) predict the extended 'C5' conformation (Fig. 1) to be the global minimum.The 'C7' conformation with an intramolecular hydrogen bond is 1.3 kJ/mol less stable than 'C5'. In contrast to the present *ab-initio* calculations all force fields predict the 'C7' conformation to be the global minimum of diglycine. The helical conformation of diglycine is not a local minimum on dzp-SCF-level. Similar results are obtained in the case of dialanine. Again the relative stability of the 'C7' conformation is significantly overestimated by the force fields.

Table 1: Relative energies[1] of the three low energy conformers of diglycine[2]

conformation[3]	SCF-dzp	AMBER	CVFF	ECEPP
C7	1.3	0.0	0.0	0.0
C5	0.0	13.8	12.0 [4]	3.4
a-helix	16.4 [4]	17.2	26.0 [4]	4.9

1) in kJ/mol; 2) see text; 3) see figure 1; 4) not a local minimum

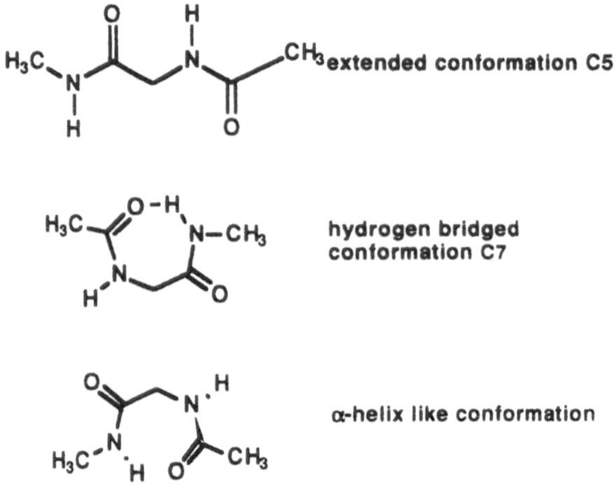

Figure 1: The three low energy conformers of Diglycine

2.2 Organolithium Compounds

Together with researchers from the plastics department (E. Hädicke, K. Knoll) and the central informatics department (L. Siggel) we have performed large scale *ab-initio* SCF and MP2 calculations on organolithium compounds, which act together with ethers as initiators for butadiene polymerization reactions. With the information on the favored conformers available through the calculations i.e. geometries, charge distributions, energies and intermolecular interactions, we proposed a new experimental protocol for the production of a plastics material.

2.3 Bithiophenes

Very recently we (together with E. Hädicke) investigated the rotational barriers in the three isomers of Dimethyl-2,2'-bithiophene(Fig 2.) .These are used as model compounds in a joint research project on nonlinear optical materials. In order to check the validity of the MNDO and PM3 parametrizations within MOPAC for this systems, we calculated the rotational energy curves on dzp-SCF level. The results show, that the MNDO curves are in much closer agreement to the *ab-initio* values, than the curves obtained with the PM3 parametrization. Even curves obtained by single-point dzp-SCF calculations based on the MNDO-geometries agree sufficiently to the geometry optimized dzp-SCF curves. This is also true for the electronic properties calculated from these geometries. This observation may help in the investigation of alkyl-substituted 2.2'-bithiophenes.

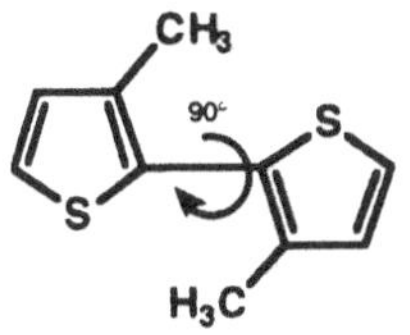

3,3′-dimethyl-2,2′-bithiophene

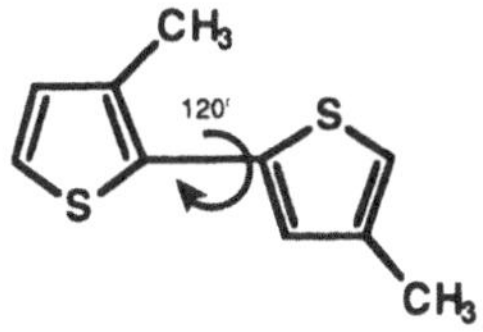

3,4′-dimethyl-2,2′-bithiophene

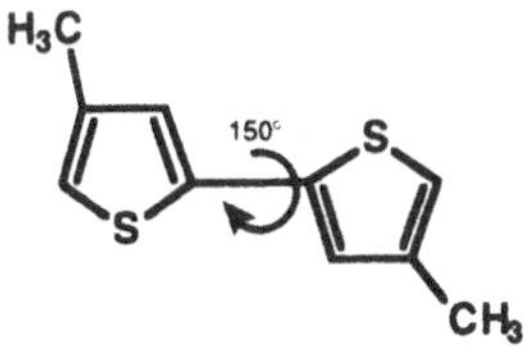

4,4′-dimethyl-2,2′-bithiophene

Figure 2: The three dimethyl-2,2′-bithiophene isomers

3. Parallel SCF

3.1 The Need for Improved Performance

The three (and also many other) examples mentioned above proved the value of quantum mechanical *ab-initio* calculations even for "industry relevant" research. However, the computational demands for sufficiently accurate calculations even on small molecules are deterringly high, as shown in table 2.

Table 2: Examples of CPU-times for *ab-initio* calculations

molecule[1]	atoms	basis-set	n_{bf}[2]	method	computer[3]	time[4]
dialanine	22	dzp	210	SCF	4D/220	7
diglycine	19	tzp	231	SCF	4D/220	10
diglycine	19	tzp	231	MP2	C210	10
bithiophene	22	dzp	246	SCF	IBM530	10

1) refer to text; 2) number of basis functions; 3) 4D/220: SGI 4D/220 GTXB workstation, C210: Convex C210 minisupercomputer, IBM530 IBM 6000/530 workstation; 4) in h

On state-of-the-art RISC cpus such as the MIPS R3000 (used in SGI workstations) or the IBM RIOS (used in the IBM 6000 workstations) one geometry optimization cycle (energy and gradient calculation) may take from a few hours to one day (24h) of cpu-time for "moderate" sized molecules. Considering 10 to 100 optimization cycles (depending on the quality of the starting geometry and the "softness" of the molecules internal degrees of freedom) it may take days to weeks to get the results. Although such computing demands are quite affordable on dedicated workstations, the results become increasingly less interesting to the research scientists the longer it takes to obtain them. However, there are several ways to improve the response time:

- Avoid costly *ab-initio* calculations alltogether.
- Use faster computers.
- Use more intelligent and efficient algorithms.
- Distribute the work to many processors.

Avoiding costly calculations by carefully chosen basis-sets and reduced but sufficient model systems is possibly the best way to overcome computer turnaround problems. Care must be taken to avoid reducing the size of the model system to the point of rendering the results meaningless. The use of faster computers (i. e. supercomputers) seems not to be cost effective for a method which still has to prove its worth. Developing more intelligent algorithms is probably the best solution over the long run. In this paper we will focus on the last point: parallelizing *ab-initio* SCF-calculations in order to distribute the workload over many, relatively cheap, processors.

3.2 General Remarks on Parallelization

Let us first recall some of the general ideas behind parallelization to obtain information about the possible speed-up and critical bottelnecks.
In principle n equal processors may perform a computational task n times faster than one single processor of the same type. Unfortunately, there are several facts in real life which significantly reduce this ideal n-fold speed-up.
First, we are concerned with the scalar bottelneck, well known as Amdahl's law in the field of vector-processing. This can be applied to parallel processing in the following way: Let t_s be the time a computational task takes on one single cpu, n the number of cpu's used and x_s the fraction of time spent in nonparallelizable parts of the computation. Then the time t_p needed to perform the task with n processors may be expressed as

(1) $$t_p = x_s * t_s + (1-x_s) * t_s/n.$$

The speed-up S is

(2) $$S = t_s/t_p = 1 / (x_s + (1-x_s)/n).$$

For an infinite number of processors an asymptotic limit is achieved for the speed-up.

(3) $$\lim S = 1 / x_s$$

Assuming a parallelization degree of 99% (i.e. $x_s = 0.01$) leads to a maximum possible speed-up of 100. Using 100 processors the speed-up will be about 50. So only one half of the power of the 100 processors would be used in this example. The degree of processor utilisation is normally expressed as the Amdahl efficency E.

(4) $\qquad E = S / n = 1 / (1 + (n-1)*x_s)$

Up to now we did not take into account delays caused by interprocessor communication, which is needed to set up the computational tasks and to exchange input data and results. Assuming that the parallelizable part is split up into n_t independent tasks, we may approximate the time t_c needed for communication as

(5) $\qquad t_c = a * n + b * n_t.$

In this formula a*n is the time needed to exchange task independet data to set up each of the n processors and $b*n_t$ accounts for the time spent to schedule the individual tasks. Although the situation is much more complex in the general practice, this approximation (5) shows the main contributions to t_c for the problems we are concerned with in this work: The communication time increases with the number of processors and the number of tasks. If we add t_c to t_p (Eqn. 1), there are two important conclusions, that can be drawn from this:

- There is an upper limit for the number of processors beyond which additional processors do not decrease but rather increase the total time.

- In order to minimize t_c one should set up as few (but at least n) tasks as possible.

While discussing Amdahl's law we have assumed that the parallelizable part of a problem may be distributed equally over all processors. This is true if the workload may be divided into n_t independent tasks of equal length and n_t is an integer multiplier of the number of processors. However, in the general case where tasks are of unpredictable and unequal length and n_t is an abitrary number, load-imbalance may become a serious performance-decreasing problem. If we have scheduled $n_t=n$ tasks in order to minimize communication overhead, the total time to reach the solution is determined by the time for the longest task. In order to obtain a good load balance one should minimize the length of the longest task by setting up a large number of short tasks of a similar a length as possible. In this case the additional time due to poor load balance is proportional to the average task length and may be estimated by

(7) $\qquad t_l = (1-x_s) * t_s / n_t.$

Now we may write down an approximate expression for the total time t_p which takes into account effects of the scalar bottleneck, communication delays and load imbalance

(8) $\qquad t_p = x_s*t_s + (1-x_s)*t_s/n + a*n + b*n_t + (1-x_s)*t_s/n_t.$

From this we can derive expressions for the optimal numbers of processors and tasks

$$(9) \qquad n = sqrt((1-x_s)*t_s/a); \qquad n_t = sqrt((1-x_s)*t_s/b)$$

and get the best possible time for t_p as

$$(10) \qquad t_p = x_s*t_s + 2*sqrt((1-x_s)*t_s) * (sqrt(a)+sqrt(b)).$$

We see that communication aspects have a strong impact to the best achievable performance.

3.3. Concepts for the Implementation of a Parallel Direct SCF on Networked Workstations

Before we apply the foundations derived in chapter 3.2 to the SCF-problem let us first recall the basic structure of a direct SCF program as depicted in fig. 3.

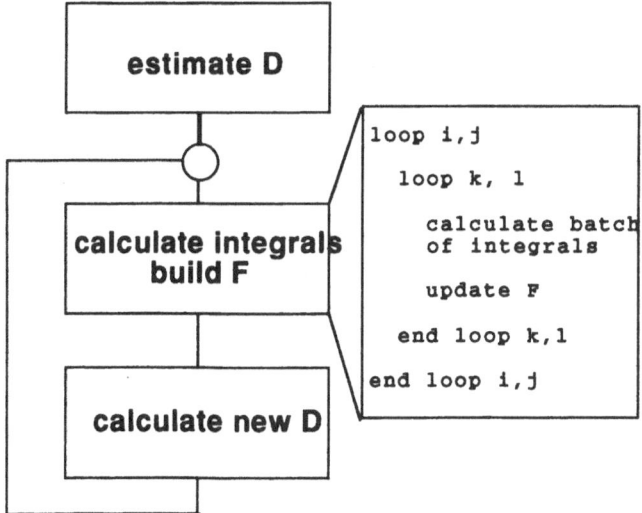

Figure 3: Scheme of a direct SCF

After an initial guess for the density matrix D has been established the fock matrix F is calculated from the density matrix and the two electron integrals I over basis functions. From the fock matrix a new density matrix is calculated and the iteration proceeds until convergency is achieved.

In a naive implementation building the fock matrix takes a time proportional to the 4th power of the number of basis functions (n_{bf}), while the time required to calculate the new density matrix increases with the 3rd power of n_{bf}. Therefore, we shall first concentrate on parallelizing the n_{bf}^4 step. Building the fock matrix is implemented as a nest of four loops over i, j, k and l (ijkl being the indices of basis function shells). Within the innermost loop the integral evaluation routines are called to obtain batches of integrals for the individual shell quadruples. Then the fock matrix is updated in an integral driven manner with an irregular reference pattern to the individual fock matrix elements. Therefore, the integral batches may be calculated independently of each other on different processors, while updating the fock matrix has to be done in scalar mode to obtain correct results. Although this approach will parallelize the bulk of the computation and requires only little local memory on the processors, performance is expected to be quite poor due to the huge (in the order of n_{bf}^4) amount of data transfer. To avoid the transfer of integrals entirely, we may also parallelize the calculation of the fock matrix, if each processor has a local copy of the density matrix and updates its private copy of the fock matrix. As an initial step the density matrix has to be transferred to each processor, then the tasks are scheduled to the processors in order to handle a specified range of the ijkl-space and finally the individual fock matrices are transferred to one processor and just summed to obtain the correct fock matrix. Now data transfer is only in the order of n_{bf}^2, which is sufficiently low compared to the cpu requirements which are of higher order. This approach allows an efficient parallelization even on a cluster of networked workstations where communication speed is relatively low, as will be shown later.

The calculation of the new density matrix, which is performed by matrix operations such as diagonalization or matrix transformations, is also parallelizable in principle. However, due to the fine granularity of loop-level parallelism, these operations cannot be parallelized efficiently on a workstation cluster at this time. This might become a problem since the computational effort to get the fock matrix is in practice only of about n_{bf}^3 and not n_{bf}^4 for large molecules. Hence, the nonparallelized part (i.e. the calculation of the new density matrix) may increase the scalar bottelneck.

For fairly large systems such as the above mentioned bithiophene systems (with 246 basis functions) runtime profiling shows, that the calculation of the fock matrix is still the dominant part. On an SGI 4D/25 with a 25 MHz R3000 MIPS cpu, building the fock matrix takes about 92% of the total time, while the calculation of the new density matrix contributes 6%. On an IBM 6000/320 with a 20 MHz RIOS cpu the situation is even better. The calculation of the new density matrix takes only about 1.3% of the total time. This is due to the superscalar architecture of the IBM RIOS chip, which performs very well on vectorisable loops.

The expected performance for a parallel direct SCF running on a cluster of such workstations can be estimated with the help of formulas (2), (8) and (9). If we assume $t_s = 3900$s per iteration, $x_s = 0.02$ (98% parallelization) and use experimental data for a = 1s and b = 0.05s we will obtain the numbers shown in column 2 (speed-up S) and 3 (efficiency E) of table 3. For the gradient calculation the parallelization degree is well over 99% and the parallel region is about a factor of 5-10 longer than in the SCF case. The numbers shown in the columns 4 (S) and 5 (E) of the table are obtained with $t_s = 35000$s, $x_s = 0.01$. The last two columns show the overall values for S and E for a complete geometry optimization cycle. Using 8 processors we can expect an overall speed-up of 6.9 at best, but we will be already satisfied if we get a factor of 6 in practice. This might bring down the times required for geometry optimizations, which may be in region of several days now, to about one night.

Table 3: Expected performance for a parallel direct SCF[1)]

number of processors	energy		gradient		total	
	S	E	S	E	S	E
2	1.9	0.96	2.0	0.98	1.9	0.97
4	3.7	0.91	3.9	0.96	3.7	0.93
6	5.2	0.82	5.6	0.94	5.4	0.90
8	6.6	0.72	7.3	0.92	6.9	0.86
16	10.8	0.67	13.4	0.83	11.8	0.74

1) see text

For the case of the gradient we already observed a speed-up of about 5 in a cluster consisting of a Convex C210, an IRIS 4D/120, an HP835 and an IRIS 4D/20 compared to the single HP835. From the theoretical model we expected a factor of 5.1. This is quite impressive, if we take into account that the total performance of this cluster is a factor 5.7 better than the single HP835.

4. Summary and Outlook

The material presented in this paper shows, that computational *ab-initio* quantum chemistry is picking up momentum in industrial research. Recent hardware developments such as cheap superscalar workstations allow cost effective computations on sufficiently large and accurate models, to help experimentalists with their everyday work. However, we still have to improve our methods in order to meet the growing demands. One possibility is to investigate the "power of parallelism" today in order to be prepared to the future developments in this field, which are expected to be impressive.

Quantum Chemical Investigations of Reactive Intermediates. Carbocations and Alkyl Radicals

Wolfram Koch

IBM Deutschland GmbH, Heidelberg Scientific Center, Institute for Supercomputing and Applied Mathematics, Tiergartenstr. 15, W-6900 Heidelberg, FRG

Abstract: Using high–level quantum chemical *ab initio* calculations employing large basis sets and including effects of electron correlation, various small carbocations and alkyl radicals are studied. Equilibrium structures, relative stabilities, harmonic frequencies and potential energy surfaces are determined and, where available, compared to experimental data. The importance of computational methods as a tool complementing experimental approaches for studying short–lived reactive intermediates is stressed.

INTRODUCTION

During the last two decades computational chemistry has evolved from an exotic branch of academic research to a recognized and important tool for the study of molecular problems of structure, stability, reaction mechanism and rates, and molecular dynamics. In particular the last few years have witnessed an enormous growth of the application of theoretical and computational methods in many areas of both academic and industrial chemical research. This development was certainly triggered and stimulated by the significant advances in computer technology. Computational chemistry has progressed hand in hand with such advances, most notably the advent of vector and parallel computers.

Theoretical methods are particularly important and useful for the study of chemical species which are not easily amenable to experiments. In these cases a computational investigation often provides answers to problems which could not be solved by only using conventional experimental methods. Moreover, the combination of reliable theoretical results with available experimental data may yield new insights in the complex chemistry of these often very unstable and short–lived species, while each approach on its own provides only restricted possibilities and a merely ambiguous description of the molecular system. In this article we will present as a case at hand of this synergistic interplay between state–of–the–art theoretical investigations and experimental results our computational results on small carbocations and alkyl radicals. Both classes of molecules are important reactive intermediates and play a key role in many reactions in organic chemistry. For example, in the S_{N1} reaction the carbocation R^+

U. Harms (Ed.)
Supercomputer and Chemistry 2
© Springer-Verlag Berlin Heidelberg 1991

generated in the first step of the reaction plays a central role in determining the products of the reaction. Detailed knowledge about the structure, the stability and the fate of the cation during its short lifetime is therefore of prime importance for an understanding of the reaction

$$RX \xrightarrow{\quad -X^- \quad} R^+ \xrightarrow{\quad +Y^- \quad} RY$$

but is usually very hard to obtain experimentally. The theoretical methods which can nowadays be employed for a study of such species have matured to a stage where accurate and reliable results can be expected. In particular the recent developments which led to methods to reliably predict vibrational (IR, Raman) and nuclear magnetic resonance (nmr) spectra offer new interfaces between theory and experiment. In the following sections we will briefly describe some of our calculations on small, but important cations ($C_3H_7^+$, $C_4H_7^+$, $C_5H_9^+$), on prototype alkyl radicals (methyl, ethyl, i–propyl, t–butyl), and finally we will present the most recent quantum chemical contribution to a "*classic*" problem of physical organic chemistry, the structure of the 2–norbornyl cation [1].

THEORETICAL METHODS

Equilibrium geometries and harmonic frequencies were obtained using standard basis sets like 6–31G*, 6–31G**, and 6–311G** at the Hartree–Fock and the correlated MP2 level of approximation. Relative stabilities have been computed at the optimized geometries employing full fourth order Møller–Plesset perturbation theory (MP4(sdtq)) for correlating the valence electrons. This level of theory should yield equilibrium geometries with bond lengths within ± 0.02–0.04Å and bond angles within ± 2–$3°$ while relative stabilities should be accurate within a few kcal/mol [2]. All calculations have been performed with the vectorized IBM version of Gaussian 86 [3] installed on an IBM 3090 with vector facility.

THE $C_3H_7^+$ POTENTIAL ENERGY SURFACE [4]

The $C_3H_7^+$ cation is the smallest carbocation stable in solution and many experimental and computational investigations have been reported. [5] However, only general features have been established, not the detailed nature of the potential energy surface. In agreement with experimental and previous theoretical results we find the 2–propyl cation to be the global minimum on the $C_3H_7^+$ potential energy surface. Protonated cyclopropane is computed to be less stable by 7.2 kcal/mol, which is in good agreement with the experimental energy difference of ca. 8 kcal/mol. The primary 1–propyl cation could not be located as an energy minimum. However, two stationary points with 1–propyl cation geometries corresponding to transition structures for hydrogen scrambling in the 2–propyl cation and the isomerization reaction from 2–propyl cation $\rightarrow$ protonated cyclopropane could be found: 19.3 and 20.5 kcal/mol above the global minimum, respectively. The computed structures offer some surprising features and deserve some comments. The 2–propyl cation prefers a *chiral* (C_2 point group) ground state, however, the methyl rotation barriers are quite small (ca. 0.5 kcal/mol). In accord with Dewar's lower level calculations[6] the energetically most favorable protonated cyclopropane geometry corresponds to an asymmetric corner protonated structure (C_s point group). The methyl rotatation *via* a symmetric corner protonated structure is virtually free. Also the edge–protonated alternative, which serves as the transition state for hydrogen scrambling in protonated cyclopropane is very close in energy, merely 1.4 kcal/mol above the most stable form.

That the *chiral* geometry of the 2–propyl cation is indeed the equilibrium structure of this ion has been established by a comparison of the ^{13}C nmr chemical shifts found experimentally in super acid media[7] with those calculated for various 2–propyl cation conformations by the IGLO method developed by Kutzelnigg and Schindler.[8] As demonstrated by the results in Table 1, only the C_2 symmetric form agrees with experiment.

Table 1: IGLO and exp.[7] ^{13}C nmr chemical shifts for 2–propyl cation conformations

Atom	exp.	C_2	C_{2v}	C_{2v}	C_s
CH	323.9	324.6	340.9	356.2	346.4
CH_3	53.8	44.5	42.7	38.2	40.3,42.7

C–H hyperconjugation is responsible for the preferred C_2 conformation of 2–propyl cation. In this conformation single hydrogens on each of the two methyl groups are aligned almost perfectly with the formally vacant orbital at the central carbon (Scheme 1).

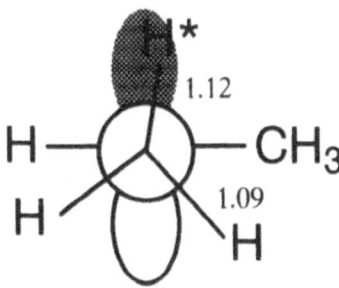

Scheme 1: Hyperconjugative interaction in the C_2 conformation of 2–propyl cation

The hyperconjugative interaction is also responsible for a lengthening of the corresponding C–H* bonds and a reduced C–C–H* angle (1.12Å and 92.7°, respectively).

Combining the results obtained for the stationary points for $C_3H_7^+$ a theoretical prediction of the potential energy surface can be constructed (Scheme 2) which is in very good agreement with the available experimental data. For example, it is experimentally observed that H–scrambling occurs slightly faster than C–scrambling in the 2–propyl cation[9] which is in line with the theoretical result of a lower transition state for the former reaction.

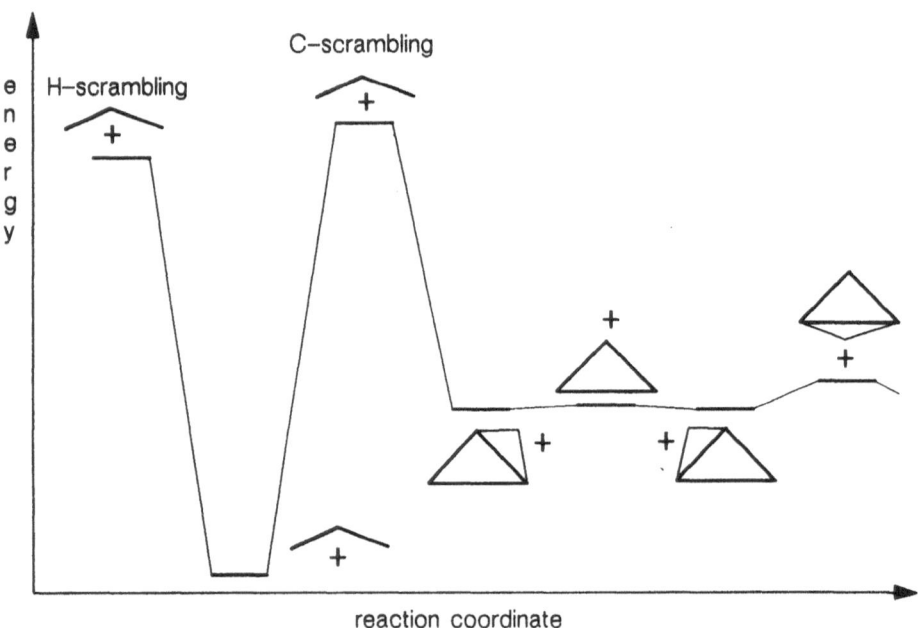

Scheme 2: $C_3H_7^+$ computed potential energy surface

THE $C_4H_7^+$ POTENTIAL ENERGY SURFACE [10]

This carbocation has received considerable interest through recent years.[5] It is experimentally well known that solvolysis of cyclobutyl, cyclopropylcarbinyl, and homoallyl derivatives generate similar product distributions (Scheme 3).

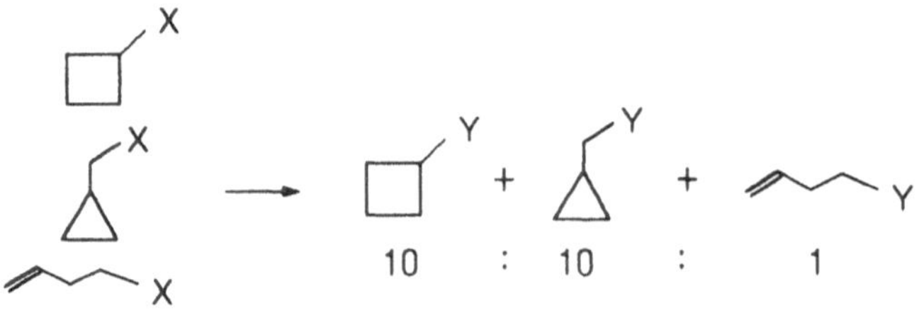

Scheme 3: $C_4H_7^+$ reactions

In solution it is now firmly established that the common $C_4H_7^+$ intermediate has more than one structure. The experimental information can best be interpreted in terms of a rapid equilibrium involving two structural isomers $C_4H_7^+$, a puckered cyclobutyl cation and cyclopropylcarbinyl cation. However, previous theoretical studies have produced ambiguous results and a complete understanding of the structures, stabilities and rearrangement pathways of the $C_4H_7^+$ cations is still lacking. Our calculations[11], which for the first time not only employed large basis sets but also included electron correlation in the geometry optimization showed that the correlation effects are of critical importance for a proper description of the $C_4H_7^+$ potential energy surface. Cyclobutyl cation, which is better described as a nonclassical, pentacoordinated bicyclobutonium ion, and bisected cyclopropylcarbinyl cation are both stable isomers (in contrast to earlier calculations[12] which computed both species as transition states, not minima) of almost identical stability, lying 9.0 kcal/mol above the global minimum, 1–methylallyl cation. The activation barrier for isomerization between the two cyclic isomers is 0.6 kcal/mol. 2–Methylallyl cation has also been found as a stable isomer, 9.5 kcal/mol above 1–methylallyl cation. All these data are consistent with the experimental information.

THE CYCLOPENTYL CATION [13]

The cyclopentyl cation is another example of the importance of high level ab initio calculations for the elucidation of structural details of carbocations. Our calculations showed that the classical C_{2v} structure of cyclopentyl cation is some 3 kcal/mol less stable than a twisted, C_2 symmetric structure with partial hydrogen bridging. Again, hyperconjugative interaction leads to a significant stabilization of the cation and determines its structure.

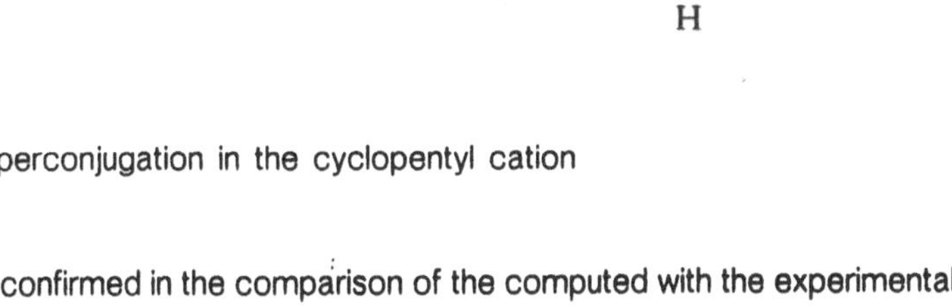

Scheme 4: Hyperconjugation in the cyclopentyl cation

This result was confirmed in the comparison of the computed with the experimentally observed ^{13}C–nmr chemical shifts. This comparison also demonstrated that for an accurate description of the equilibrium structure of the cyclopentyl cation electron correlation have to be taken into account. At the Hartree–Fock level the degree of hydrogen bridging is severely underestimated and only the ^{13}C–nmr chemical shifts computed at the MP2/6–31G** optimized geometry show good agreement with the experimental data.

ALKYL RADICALS [14]

Alkyl radicals are another very important class of reactive intermediates. They play an essential role not only in the production but also in the degradation of polymers, thus determining the stability of many materials such as coatings and lubricants towards heat, light, and high energy radiation. An accurate knowledge of the equilibrium structures and the vibrational frequencies is a prerequisite for a detailed understanding of the chemistry of these species. In our calculations we chose the series of radicals, methyl (CH_3), ethyl (C_2H_5), isopropyl (($CH_3)_2CH$), and tertiary butyl (($CH_3)_3C$), because they are the prototypes for primary, secondary and tertiary radical centers. The geometry optimizations indicate that for alkyl radicals, which, just as carbocations can be regarded as electron deficient systems, hyperconjugation is also an important, structure determining factor. The typical features of hyperconjugation, i.e., elongated β–CH bonds parallel to the electron deficient center can be found in all species studied (with the obvious exception of the methyl radical). However, this effect is of course not as pronounced as in similar carbocationic systems. For example, in the t–butyl radical the two CH bond lengths differ by 0.008Å (1.102 and 1.094Å), while in the corresponding t–butyl cation the difference amounts to 0.019 (1.103 and 1.084Å). The non–planarity of the alkyl radicals (with the exception of CH_3, which is planar) can also be understood in terms of hyperconjugative interaction. Besides the equilibrium geometries we were interested in the vibrational spectra of these species, which are, e.g., very important for the calculation of entropies using third law calculations. Due to the particularities described above in the geometries, an alkyl radical has specific vibrational modes and can be considered as a functional group. The most important of these alkyl radical specific modes are the β–CH stretch and the pyramidal bending motion of the radical center.

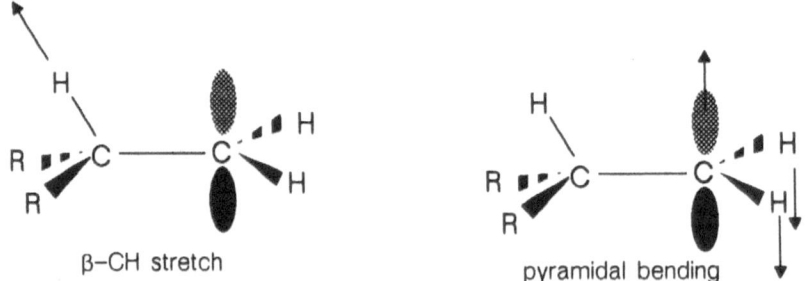

β–CH stretch   pyramidal bending

Scheme 5: Characteristic modes in alkyl radicals

The β–CH stretch modes are found experimentally as well as computationally in spectral regions where organic systems containing only carbon and hydrogen do not have any absorbtion (exptl. 2800–2840 cm^{-1}). The theoretically predicted spectra for the pyramidal bending motion also reproduce the experimental ones excellently for all systems, but t–butyl. No experimental band above 200 cm^{-1} was observed, although the calculations predict an intense band at around 280 cm^{-1}. Consequently, either theory may be predicting too high an intensity for this mode or the computed band center is off. The latter situation, i.e., that the experimental frequency is below 200 cm^{-1} it would be outside the detection of the IR experiments. This is certainly a possibility considering that the motion is very anharmonic and involves a large amplitude. In general, excellent agreement was found between the theoretical and observed IR spectra of the alkyl radicals. This provides credence for the computed radical structures, and furthermore provides additional information for the interpretation of the geometries and vibrational spectra.

THE STRUCTURE OF THE 2–NORBORNYL CATION [15]

The equilibrium structure of the 2–norbornyl cation is possibly one of the most widely debated issues in physical–organic chemistry [16]. Many experiments demonstrate that the 2–norbornyl cation seems to profit from some kind of stabilization not present in other similar cations. This is illustrated by comparing the rates of the two solvolysis reactions in Scheme 6. Solvolysis of 2–tosyl–norbornane to give the 2–norbornyl cation is about 400 times faster than the analogous reaction of tosyl–cyclopentyl to give the cyclopentyl cation.

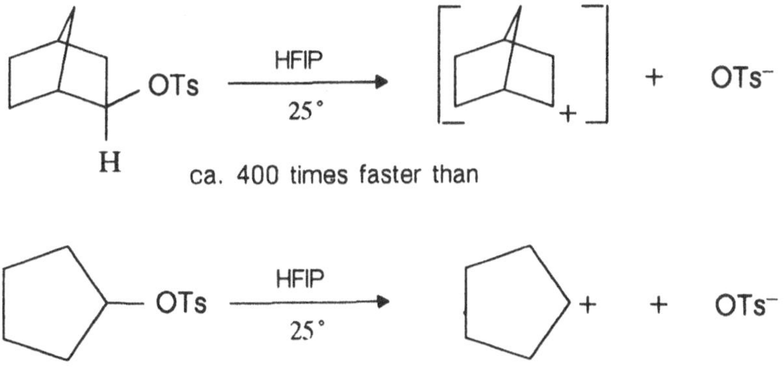

Scheme 6: Solvolysis reaction of 2–tosyl–norbornane and tosyl–cyclopentane in hexafluoroisopropanol (HFIP) as solvent

What is the reason for this surprising stability of the 2–norbornyl cation? Two explanations are discussed in the literature (Scheme 7): A rapid equilibrium between the two equivalent structures A and B, with a symmetrical structure (C) as the saddle point, or only one symmetrical structure (C) with a 3–center–2–electron bond, which violates the classical bonding rules

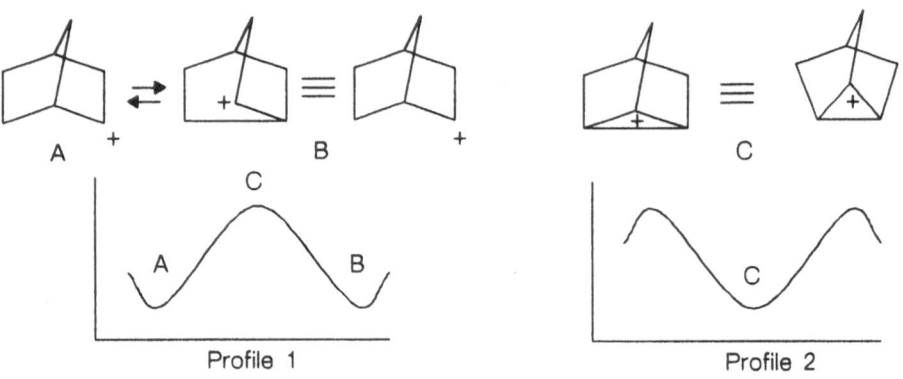

Scheme 7: Possible 2-norbornyl structures and energy profiles

and is therefore termed a *non-classical* cation. Most of the experimental information point to a symmetrically bridged structure C as the only minimum, although some uncertainty still remains. Theoretical calculations are in principle ideally suited to give an unambiguous answer to this problem. Computationally, it is sufficient to show whether C represents a saddle point (as in profile 1) or a minimum (as in profile 2) on the potential energy hypersurface. However, such calculations on an adequate theoretical level for a molecule of that size are very time consuming and beyond the computational capabilities of most research groups. Also, for the molecule at hand the high level calculations required have not yet been carried out, while calculations at lower levels of approximation yield ambiguous results. Thus, the final theoretical proof whether the non-classical 2-norbornyl cation is a potential minimum is still missing. We decided to tackle this problem by performing a systematic investigation using high levels of *ab initio* theory. The geometry of the non-classical ion (C) has been optimized using several standard basis sets of increasing quality: 3-21G, 4-31G, 6-31G, 6-311G, 4-31G*, 6-31G*, 6-31G**, and 6-311G*. The first four basis sets consist only of s- and p-type functions, while the remaining four are so-called polarized basis sets, which include additional d-type functions on carbon, and for 6-31G** also p-type functions on hydrogen. At the optimized geometries the force constant matrix (i.e., the second derivatives of the energy with respect to the nuclear coordinates) was computed analytically and from this the harmonic frequencies were constructed. A minimum is characterized by only real frequencies (and a positive definite force constant matrix), while a saddle point has one imaginary frequency (and a negative eigenvalue in the force constant matrix). The absolute value of the critical frequency is an indicator for the steepness of the potential energy hypersurface along the corresponding normal mode. The values of the lowest frequency, which corresponds to the symmetry breaking vibrational mode moving the system towards the classical structure (A,B), computed with the various basis sets, are shown in Table 2. It can be seen that with unpolarized basis sets the lowest frequency is imaginary and C represents a saddle point. Once the basis set includes

Table 2: 2–Norbornyl cation, lowest frequency [cm^{-1}]

Basis	3–21G	4–31G	6–31G	6–311G	4–31G*	6–31G*	6–31G**	6–311G*
ν_{min}	137i	136i	154i	182i	70	53	69	116

polarization functions, the frequency turns real and its absolute value increases with increasing size of the basis. Thus we conclude that the non–classical structure C is a minimum on the Hartree–Fock potential energy surface of the 2–norbornyl cation. However, electron correlation has not been included in these calculations. therefore we repeated the optimization and force constant matrix determination (by numerically differentiating the analytically obtained gradients) using second order Møller–Plesset perturbation theory and a 6–31G* basis set. The vibrational mode that moves the system towards the classical structure is now the second lowest (not the lowest as before) mode at 291 cm^{-1}, compared to 53 cm^{-1} at the Hartree–Fock level with the same basis set. Thus, electron correlation makes the non–classical well considerably steeper, increasing the vibrational frequency by 238 cm^{-1}. The conclusion that the equilibrium gas phase structure of the 2–norbornyl cation is indeed non–classical is further supported by a comparison of our computed MP2/6–31G* harmonic frequencies with the IR spectrum recently obtained experimentally by Sunko et al. using a new technique [17].

CONCLUSIONS

The examples described in this article have shown that "state–of–the–art" quantum chemical ab initio calculations are indeed a very helpful tool for the study of short–lived reactive intermediates. It becomes particularly evident that for carbocations non–classical structures are more the rule than the exception. Hyperconjugation is an important source of stabilization for carbocations and is also operative in alkyl radicals. The availability of new interfaces between theory and experiment like nmr chemical shifts or harmonic frequencies offers new and sensitive probes for structure identification.

ACKNOWLEDGMENTS

I wish to thank all the colleagues who were involved in the various projects described above. In particular I am grateful to D.J. DeFrees (San Jose), B. Liu (San Jose), J. Pacansky (San Jose), P.v.R. Schleyer (Erlangen), and D.E. Sunko (Zagreb). Excellent services, provided by the computing centers at the IBM Almaden Research Center, San Jose (CA), the IBM Heidelberg Scientific Center and the IBM Düsseldorf facility are gratefully acknowledged. I also thank C. Thümmler for carefully reading the manuscript.

REFERENCES

[1] Due to space limitations, the discussion of the various investigations will be kept very brief. For a detailed description and an exhaustive list of references, in particular for the experimental background, the reader is referred to the original papers.

[2] For a description of the theoretical methods and a detailed discussion of their accuracy see: W.J. Hehre, L. Radom, P.v.R. Schleyer, J.A. Pople, *Ab initio Molecular Orbital Theory*, Wiley–Intescience, New York (1986).

[3] Gaussian 86: M.J. Frisch, J.S. Binkley, H.B. Schlegel, K. Raghavachari, C.F. Melius, R.L. Martin, J.J.P. Stewart, F.W. Bobrowicz, C.W. Rohlfing, L.R. Kahn, D.J. DeFrees, R. Seeg er, R.A. Whiteside, D.J. Fox, E.M. Fluder, J.A. Pople, Carnegie–Mellon Quantum Chemistry Publishing Unit, Pittsburgh, PA (1984)

[4] (a) W. Koch, B. Liu, P.v.R. Schleyer, *J. Am. Chem. Soc.*, **111**, 3479 (1989), (b) P.v.R. Schleyer, W. Koch, B. Liu, U. Fleischer, *J. Chem. Soc., Chem. Commun.* 1098 (1989)

[5] See e.g. P. Vogel, *Carbocation Chemistry*, Elsevier, Amsterdam (1985)

[6] M.J.S. Dewar, E.A. Healy, J.M. Ruiz, *J. Chem. Soc., Chem. Commun.* 943 (1987)

[7] G.A. Olah, D.J. Donovan, *J. Am. Chem. Soc.*, **99**, 5026 (1977)

[8] (a) M. Schindler, *J. Am. Chem. Soc.*, **109**, 1020 (1987), (b) W. Kutzelnigg, *Isr. J. Chem.*, **19**, 173 (1980), (c) M. Schindler, W. Kutzelnigg, *J. Chem. Phys.*, **16**, 1919 (1982)

[9] M. Saunders, P. Vogel, E.L. Hagen, J. Rosenfeld, *Acc. Chem. Res.*, **6**, 53 (1973). See also ref. 5a

[10] W. Koch, B. Liu, D.J. DeFrees , *J. Am. Chem. Soc.*, **110**, 7325 (1988).

[11] For a similar theoretical study, which appeared parallel to ours, see: M. Saunders, K.E. Laidig, K.B. Wiberg, P.v.R. Schleyer, *J. Am. Chem. Soc.*, **110**, 7652 (1988)

[12] M.L. McKee, *J. Phys. Chem.*, **90**, 4908 (1986)

[13] P.v.R. Schleyer, J.W.d.M. Carneiro, W. Koch, K. Raghavachari, *J. Am. Chem. Soc.*, **111**, 5475 (1989)

[14] J. Pacansky, W. Koch, M.D. Miller, *J. Am. Chem. Soc.*, in press

[15] (a) W. Koch, B. Liu, D.J. DeFrees, *J. Am. Chem. Soc.*, **111**, 1527 (1989), (b) W. Koch, B. Liu, D.J. DeFrees, D.E. Sunko, H. Vancik, *Angew. Chem.* **102**, 198 (1990)

[16] Recent reviews: C.A. Grob, *Acc. Chem. Res.*, **16**, 426 (1983), H.C. Brown, *ibid*, 432, G.A. Olah, G.K.S. Prakash, M. Saunders, *ibid*, 440, C. Walling, *ibid*, 448, D. Lenoir, Y. Apeloig, D. Arad, P.v.R. Schleyer, *J. Org. Chem.* **53**, 661 (1988)

[17] H. Vancik, D.E. Sunko, *J. Am. Chem. Soc.*, **111**, 3742 (1989) and ref. 15b

Long Time Dynamics of Proteins:
an off-Lattice Monte Carlo Method

E. W. Knapp and A. Irgens-Defregger

Fachbereich Chemie, Institut für Kristallographie, Freie Universität Berlin,
Takustr. 6, W-1000 Berlin 33, FRG

Physik Department, Technische Universität München, W-8046 Garching, FRG

Abstract

A new method is proposed to perform computer simulations of protein dynamics in the
long time regime. The method is based on a Monte Carlo technique. The only molecular
degrees of freedom considered are bond rotations. All other degrees of freedom including
the amide plane torsions are kept rigid. These constraints can approximately account for
all interactions related to the chemical bonding. To provide a first test of the method, the
non–bonded interactions are turned off in the present application.

An individual Monte Carlo step attempts a local conformational change in a small window
of the protein backbone containing 3 residues in consecutive order. The residues outside of
the window remain invariant. Ergodicity of the model is demonstrated by comparing exact
thermodynamic averages with values obtained from computer simulation data involving
2×10^6 scans of the window along the protein backbone. By relating time with the number
of scans of the window along the protein backbone one can evaluate time correlation
functions. The end–to–end distance autocorrelation function decays in about 10^3–10^4 scans
of the window algorithm. Even in the absence of non–bonded interactions the time decay
follows a stretched exponential Kohlrausch decay law.

1. INTRODUCTION

Due to hidden complexities chemical systems can behave very irregular. It is then difficult
to extract the relevant components of the system which serve as a guideline to a simplified
numerical or even analytical treatment. The goal of such an analysis is to understand the
behaviour of the chemical system on a microscopic basis of description. Often this goal can
only be reached by first performing intensive computer simulations of the system under

U. Harms (Ed.)
Supercomputer and Chemistry 2
© Springer-Verlag Berlin Heidelberg 1991

study. A subsequent analysis of the data of the computer simulation can reveal the complex nature of the considered chemical system. Such studies can only be performed if large amounts of CPU time on supercomputers or workstations with high performance are available. This is also relevant for complex biological systems. Hence one can argue that in the future major demands on CPU time will come from the fields of chemistry and biology.

Experiments cannot simply be replaced by computer simulations. In the contrary experimental data provide a necessary control on the validity of simulation data. On the other hand data from computer simulations can be much more detailed than experimental data. Hence they can yield the necessary link to obtain a better understanding of the considered system. Furthermore computer simulations can replace experiments for routine applications which have been checked under similar conditions.

Protein molecules are complex biological systems. Globular proteins represent an important class of proteins [1-3]. Enzymes which function as chemical factories in the living cell are globular proteins. There are two main problems related with globular proteins. The first problem is to understand protein function on a microscopic basis of description. This requires knowledge of three dimensional (3D) structure and dynamical fluctuations of the protein. The second problem is to unterstand the mechanism of protein folding [4-6]. It can also be phrased in how does the protein fold in its distinct native structure. Apparently the second problem involves long time dynamics of the protein.

Proteins are very rich of dynamical phenomena [7,8]. Their time scale extends from 10^{-14} s for intramolecular vibrations to $10^{-3} - 10^{+3}$ s refering to protein folding processes. Within this time interval one observes fluctuations of side chains and loops, rigid body motions of helices, domains and other subunits, proton and water diffusion. Some of these processes are closely related to protein function. Computer simulation of protein dynamics is now becoming a universal tool to study these problems on an atomistic level of description[7-13].

The conventional method of computer simulation of protein dynamics is based on the solution of the classical equations of motion for each atom. Due to the very fast intramolecular vibrational degrees of freedom the time step of propagating the solution is normally 1 fs. This limits the present day application of this method to the calculation of a small number of trajectories which extend up to the nanosecond time regime. However, many functionally important motions (fims) in proteins occur in the micro– to millisecond time regime[14,15]. By using holonomic constraints one can eliminate stiff degrees of freedom as for instance the high frequency bond stretching vibrations of OH and NH[16-20]. This allows to increase the elementary time step of propagation by a factor of 2 with a corresponding speed up in execution. However, it does not pay off to eliminate also the low frequency bending vibrations since the algorithm used to fulfill the constraints needs more computer time than can be saved by increasing the time step of propagation[19].

Two different Monte Carlo methods have been applied to simulate protein structure[21,22] and dynamics[23]. The protein dynamics study[23] using the Monte Carlo method does not apply constraints on bond stretching and bond angle bending. Individual Monte Carlo steps involve a small but arbitrary displacement of individual atoms of the protein. Very often such a motion gives rise to considerable bond stretching or bond angle bending. Then the new protein conformation has a high energy and enters statistical averages with a low weight. This renders the method less efficient than conventional computer simulation of molecular dynamics. In the other approach[21,22] the monomer units are placed on a lattice whose grid points correspond to proper values of bond lengths and bond angles. The underlying model is built with non–specific point–like monomer units. It cannot be generalized to a protein model with a realistic backbone and individually different residues.

In polymer science Monte Carlo methods are often used. Normally one investigates equilibrium properties of polymers[23-25]. But this technique is also used to simulate the dynamics of polymers[24-26]. In these applications the monomer units of a polymer are also treated as structureless point–like particles placed on the grid points of a regular lattice (on–lattice polymer model). The geometry of the lattice refers to the bond lengths and bond angles of the polymer model. Polyethylene with tetrahedal bond angles (109^0) for instance corresponds to the diamond lattice.

To employ this technique for dynamics simulation the individual Monte Carlo steps must resemble realistic dynamical fluctuations. A simple Monte Carlo method employs the so called pivot or wiggle algorithm[25] where the individual Monte Carlo step is a discrete rotation around a single randomly chosen bond. This method involves large displacements of monomer units which are far away from the rotation axis. Therefore such Monte Carlo steps cannot be related to realistic dynamical fluctuations of the polymer model. Furthermore for very densely coiled polymer conformations this sampling method becomes inefficient. In many cases monomer units overlap strongly after an attempted conformational change such that the new conformation is not accepted. Another method is based on the reptation model[28] where a monomer unit is cut at one end and pasted at the other end in an arbitrary orientation. This method is very efficient for equilibrium properties. But it cannot easily be applied to heterogeneous polymers with different sidechains like a protein. Local conformational changes involving only a small number of monomer units from an on-lattice polymer model can be obtained by using kink jump or crankshaft motions[29,30]. An analogous algorithm for off–lattice polymer models with fixed bond lengths and bond angles involves rotations around 6 or more bonds[31,32] as will be demonstrated in section 3. Such an off-lattice algorithm is complicated but rather close to realistic motions.

For on–lattice simulations the chemical structure of a polymer model is stable since the geometry of the lattice accounts for bond lengths and bond angles of the polymer. This is not the case for off–lattice simulations of polymers. To avoid numerical instabilities special precautions must be taken. To the best of our knowledge off–lattice polymer models have not been applied for an intensive simulation of polymer dynamics yet.

The type of dynamics which can be handled by a Monte Carlo method is a stochastic diffusion–like motion on the potential energy surface in the coordinate space. The dimension of this space is given by the number of explicitly considered degrees of freedom. The method does not account for kinetic energy contributions as does the dynamics based on Newton's equations of motion. Hence small amplitude atomic motions due to high frequency harmonic vibrational modes cannot be described correctly in contrast to large amplitude drift motions of molecular groups. However, the high frequency vibrational modes can serve as a heat bath for the explicitly considered relevant degrees of freedom, i. e. the bond rotations.

A major problem of this approach is the lack of knowledge of the time unit refering to a single Monte Carlo step. To solve this problem one can make an adjustment to the characteristic time scale of a key experiment. This can for instance be Mössbauer spectroscopy, which provides detailed information on the dynamics of iron containing proteins in the time regime from 10^{-9} to 10^{-7}s [14,33]. Another possibility to gauge the time unit of Monte Carlo steps is the comparison with data from a bench mark calculation, where protein dynamics is simulated by solving Newton's equations of motion at very long times.

What gain can be expected by performing a protein dynamics simulation with a Monte Carlo method, where all stiff degrees of freedom (bond lengths and bond angles) are kept rigid? Fast vibrations involving large restoring forces and small amplitude motions are eliminated. They can serve as a heat bath for the rotational degrees of freedom. The remaining degrees of freedom give rise to coherent motions of larger molecular groups. The eliminated vibrational degrees of freedom impose a strong damping mechanism which renders the motion of the molecular groups diffusion–like. These effects result in a time unit of the elementary Monte Carlo step which is probably several orders of magnitude larger than the elementary time step appropriate to propagate Newton's equations of motion. After turning on the non–bonded interactions a single Monte Carlo step requires a considerable fraction of computer time needed for a single time step to propagate Newtons equations of motion (see also section 7). All together we consider this method to be much more efficient and suitable for the simulation of protein dynamics in the long time regime.

In this article we employ a Monte Carlo method to simulate protein dynamics. Thereby we are constraining all bond lengths and bond angles as well as the torsion angle of the amide planes. For test purposes we start with computer simulations where the non–bonded interactions are turned off. A first account on this method has been published recently[34].

We beginn with a short description of the conventional method of computer simulation of protein dynamics (section 2). In section 3 the idealized model for the protein backbone is introduced. The elementary Monte Carlo step for local conformational changes of the backbone is described in section 4. In section 5 we present results on equilibrium properties, in section 6 time correlation functions of the dynamics of a protein backbone are discussed. It provides a check on the proper statistical and dynamical behaviour of the algorithm. Future developments of this approach are discussed in section 7.

2. THE CONVENTIONAL METHOD OF COMPUTER SIMULATION OF PROTEIN DYNAMICS

The method of computer simulation is based on the solution of Newton's equations of motion for all N atoms (i) of the protein

$$m_i \frac{d^2}{dt^2} \vec{r}_i = - grad_i \, E(\vec{r}_1, \vec{r}_2, \dots, \vec{r}_N) , \qquad i = 1, 2, \dots, N \tag{1}$$

The energy function is split into 3 different contributions

$$E = E_{bond} + E_{nbond} + E_c \tag{2}$$

The internal energy E_{bond} accounts for the chemical bonding and geometry of the different molecular units. It consitst of 4 terms

$$E_{bond} = \sum_{[b]} k_b \, (\vec{r}_b - \vec{r}_{0b})^2 + \sum_{[\theta]} k_\theta \, (\theta - \theta_0)^2$$

$$+ \sum_{[\phi]} k_\phi \, (1 - \cos(n\phi)) + \sum_{[\omega]} k_\omega \, (\omega - \omega_0)^2 \tag{3}$$

refering to bond stretching, bond angles, dihedral angles and improper torsions, respectively. The interactions of non–bonded atoms E_{nbond} account for electrostatic and Van der Waals interactions

$$E_{nbond} = \sum_{i<j} \frac{q_i \, q_j}{\epsilon \, r_{ij}} + \sum_{i<j} \left[\frac{A_{ij}}{r_{ij}^{12}} - \frac{B_{ij}}{r_{ij}^{6}} \right] \tag{4}$$

where q_i is the partial charge at atom i and ϵ an effective dielectric constant.

Hydrogen bonds and the interaction of non–bonded atoms are both of electrostatic origin. The parameters of the non–bonded energy terms (4) are typically defined such that the effect of hydrogen bonding is included. An explicit evaluation of hydrogen bond interactions would require more computer time. The third contribution to the energy function (2) is used to stabilize certain geometries. It can for instance be used to fix an atom coordinate $\vec{r}_i$ to its x–ray structure position $\vec{r}_{xi}$ by using harmonic constraints $E_c = k_c(\vec{r}_i - \vec{r}_{xi})^2$.

The propagation of the equations of motion (1) in time can be made with the Verlet algorithm[19]

$$\vec{r}_i(t + \Delta t) = 2\vec{r}_i(t) - \vec{r}_i(t - \Delta t) - \frac{(\Delta t)^2}{m_i} grad_i\, E(\vec{r}) \tag{5}$$

which requires the knowledge of 2 subsequent sets of atomic coordinates. The propagation is started by using one coordinate set from the x–ray structure analysis. The second coordinate set is calculated from initial velocities which are taken at random from a Maxwell distribution of velocities for a given temperature.

Before one can use the coordinates of the x–ray structure hydrogen atoms have to be added. This is done for the polar hydrogens only. To save computer time the nonpolar hydrogen atoms are taken into account by using suitably defined extended atoms as for instance for the methyl group. The high frequency motions can be frozen by constraints. This method is normally applied for the bond stretching of the H-atoms by using the SHAKE algorithm[19]. It allows to save computer time by increasing the time increment Δt from typically 1 fs to 2 fs.

The added hydrogens and minor inconsistencies between the x–ray structure and the energy function (1) lead to stress in the initial structure. The stress is relaxed by minimizing the potential energy. The potential energy of the resulting protein structure does not yet correspond to a particular temperature. This is accomplished by a dynamics simulation over several 10 ps where the velocities are adjusted several times. The subsequent portion of the computer simulation can then be taken for the analysis.

3. THE PROTEIN BACKBONE MODEL

Now we introduce the idealized protein backbone model which is used for the Monte Carlo method to simulate protein dynamics. In this model the majority of the degrees of freedom are frozen. This is justified since normally bond length variations of the protein backbone do not exceed 0.05Å and bond angle variations are below 5 degrees. Furthermore also the amide planes defined by the four consecutive backbone atoms $C_j(\alpha) - C_j - N_{j+1} - C_{j+1}(\alpha)$ remain planar with torsion angle deviations of less than 5 degrees[1,2]. These degrees of freedom are kept rigid. Hence the only degrees of freedom left for the protein backbone motion are the rotations around the $C_j(\alpha) - N_j$ bonds (rotation axis $\vec{n}_j$) and around the $C_j(\alpha) - C_j$ bonds (rotation axis $\vec{c}_j$).

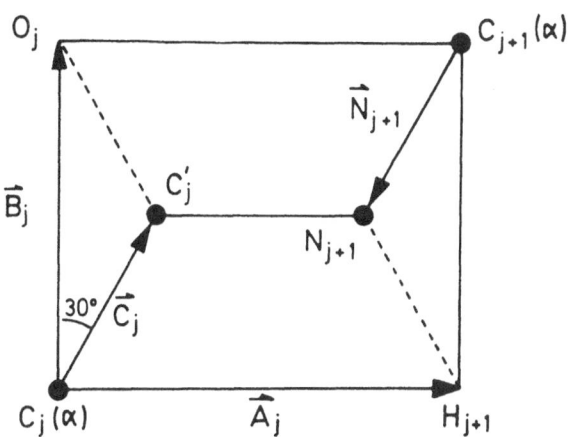

Fig. 1: A cartoon of the jth amide plane used for test simulations of the protein backbone is depicted. The rectangle is spanned by the pair of orthogonal vectors $\vec{A}_j$ and $\vec{B}_j$. The rotation axes $\vec{C}_j$ and $\vec{N}_{j+1}$ point from atom $C_j(\alpha)$ to C_j' and from atom $C_{j+1}(\alpha)$ to N_{j+1} respectively. All atoms including the hydrogen atom H_{j+1} and the oxygene atom O_j are in the amide plane, but the latter atoms are not exactly in the corners of the regular rectangle. Capital letters refer to non-normalized vectors. The corresponding vectors of unit length are denoted by lower case letters.

As a test case for an algorithm of conformational changes we consider a simplified version of the amide plane which constitutes the repeat unit of the protein backbone. It has the appearance of a letter envelope with a rectangular shape (see Fig. 1). The $C(\alpha)$ atoms are situated at diagonal corners of the rectangle. The rotation axes point from the $C(\alpha)$ atoms towards the N and C atoms in the central part of the amide planes. They are given as linear combinations of the unit vectors $\vec{a}_j = \vec{A}_j / |\vec{A}_j|$ and $\vec{b}_j = \vec{B}_j / |\vec{B}_j|$, which span the jth amide plane as follows

$$\vec{n}_{j+1} = \alpha_j^n \, \vec{a}_j + \beta_j^n \, \vec{b}_j$$

$$\vec{c}_j = \alpha_j^c \, \vec{a}_j + \beta_j^c \, \vec{b}_j, \qquad\qquad \vec{a}_j \cdot \vec{b}_j = 0 \qquad (6)$$

where

$$\alpha_j^c = -\alpha_j^n = \tfrac{1}{2}$$

$$\beta_j^c = -\beta_j^n = \tfrac{\sqrt{3}}{2}. \qquad (7)$$

For a more realistic shape of the amide planes the coefficients α and β assume slightly different values. Note that all vectors denoted by lower case symbols are of unit length.

The model amide planes are pasted together by requiring that the angle between the rotation axes $\vec{n}_j$ and $\vec{c}_j$ at the $C_j(\alpha)$ atom is tetrahedral i. e.

$$\vec{n}_j \cdot \vec{c}_j = -\tfrac{1}{3}. \qquad (8)$$

This can be fulfilled by adding the j+1 rectangle in the plane of the j rectangle as follows

$$\vec{a}_{j+1} = \tfrac{1}{3}\vec{a}_j - \tfrac{2\sqrt{3}}{3}\vec{b}_j$$

$$\vec{b}_{j+1} = \tfrac{2\sqrt{3}}{3}\vec{a}_j + \tfrac{1}{3}\vec{b}_j. \qquad (9)$$

This orientation corresponds to a trans–cis configuration with torsion angles ($C_{j-1} - N_j - C_j(\alpha) - C_j$) $\phi_j = 180^0$ and ($N_j - C_j(\alpha) - C_j - N_{j+1}$) $\psi_j = 0^0$. A random configuration of the protein backbone is obtained by adding a new amide plane first in the plane of the preceding amide plane. Then the new amide plane is rotated by random angles about the two rotation axes at the common $C(\alpha)$ atom. The application of the rotation operations is facilitated by the fact that all rotations interchange since the rotation axes are not space fixed but are body fixed.

Without precautions the above described algorithm which produces a random protein backbone conformation is unstable. Each time the freshly pasted amide plane j+1 is rotated around the two axes $\vec{n}_{j+1}$ and $\vec{c}_{j+1}$, about one significant digit is lost. As a result the basis vectors $\vec{a}_{j+1}$ and $\vec{b}_{j+1}$ spanning the new amide plane are no longer exactly orthogonal. The validity of the recursion scheme is based on the orthogonality of the vector pair $\vec{a}_j$ and $\vec{b}_j$ on the right hand side of the Eqs. (9). Hence after a few amide planes have been pasted at the protein backbone according to the recursion relation (9) all significant digits are lost. This problem can be avoided by applying infinitesimal rotations at the vectors $\vec{a}_{j+1}$ and $\vec{b}_{j+1}$ each time a new $(j+1)^{\text{th}}$ amide plane has been pasted. The infinitesimal rotations with axis $\vec{a}_{j+1} \times \vec{b}_{j+1}$ are applied such that they reinstall orthogonality but do not change the orientation of the rotation axes $\vec{n}_{j+1}$ and $\vec{c}_{j+2}$ situated in the freshly pasted amide plane.

4. THE WINDOW ALGORITHM

We seek an algorithm to perform local conformational changes, where 3 consecutive amide planes of a protein backbone are moved without a change in the other parts of the protein. For this purpose we consider a piece of 5 amide planes from the protein sequence. Only the 3 amide planes in the window are supposed to undergo a conformational change (Fig. 2).

window algorithm of conformational changes

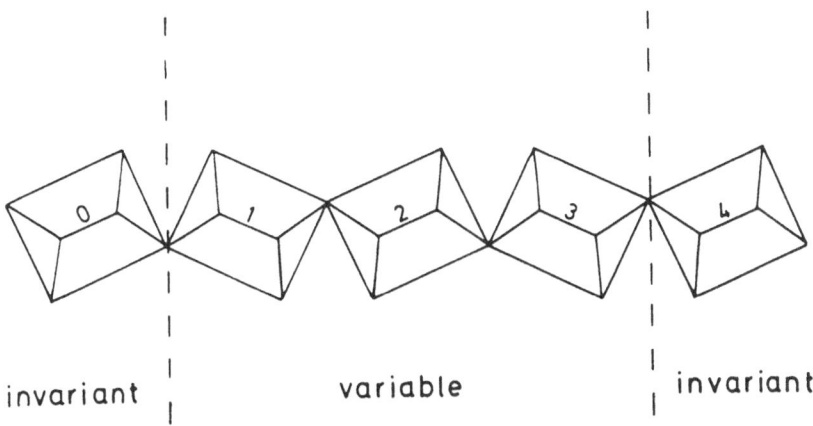

invariant   variable   invariant

Fig. 2: A segment of the idealized protein backbone is depicted. The amide planes number 1,2 and 3 in the window are considered to undergo a conformational change which leaves the 0th and 4th amide planes invariant. The tetrahedral angle between the vectors $\vec{N}_j$ and $\vec{C}_j$ remains at 109°.

The first (0) and the last (4) amide plane as well as the rest of the protein sequence remains invariant. There is a total of 8 rotational degrees of freedom involved in a conformational change of the 3 amide planes in the window. The corresponding rotation axes of unit length are $\vec{n}_j$, $\vec{c}_j$, $j = 1,2,3,4$. To count the number of conditions which must be fulfilled to keep the remaining part of the protein backbone invariant, it is convenient to assume that all rotations are applied towards the amide planes with increasing index. With this convention the orientation of the 0th amide plane remains invariant without constraints. Conformations which are obtained with other conventions may differ only by a rotation of the whole protein backbone. To keep the 4th amide plane outside of the window invariant one has to fulfill a total of 6 conditions, 3 for the absolute position and 3 for the orientation of the 4th amide plane in agreement with Refs. 31, 32. Hence 2 of the available 8 rotations can be performed with arbitrary angles and the remaining 6 unknown rotation angles are determinded by fulfilling the 6 constraining conditions. In Ref. 31 it was shown that for the case of a window with only 2 amide planes and 6 unknown rotations the problem can be simplified to a single algebraic equation by eliminating the other unknowns with a series of substitutions. This procedure should in principle also be possible for the more general cases of a window with 3 or more amide planes. In that case each choice of the arbitrary rotation axes requires its own specific scheme of eliminating the unknown rotation angles. For a window with 3 amide planes one has 2 arbitrary rotations which can be at $\begin{bmatrix} 8 \\ 2 \end{bmatrix}$ = 28 different positions. These combinatorial problems are one reason that this method has not been taken up yet.

The 6 conditions will now be formulated for a window with 3 amide planes. For that purpose we denote the rotations around the axes $\vec{c}_j$ and $\vec{n}_j$ with angles ϕ_j and ψ_j by the operators $\underline{C}_j$ and $\underline{N}_j$ respectively. The first 3 conditions can be fulfilled by requiring that the sum $\Sigma \vec{R}_j$ of the diagonal vectors $\vec{R}_j = \vec{A}_j + \vec{B}_j$ of the 3 amide planes of the window does not change after a conformational transition has occured

$$\sum_{j=1}^{3} (\vec{R}_j - \vec{R}_j^{(0)}) = 0 \tag{10}$$

where the new diagonal vectors are given by

$$\vec{R}_j = \prod_{k=1}^{j} \underline{C}_k \, \underline{N}_k \, \vec{R}_j^{(0)} \; . \tag{11}$$

The superscript zero denotes the structural quantities from the old conformation. Without a superscript they refer to the new conformation. To keep the orientation of the 4th amide plane invariant we require that the product of the 8 rotations is the identity operator

$$\prod_{j=1}^{4} \underline{C}_j \, \underline{N}_j = \underline{1} \; , \tag{12}$$

where $\quad (\underline{1})_{ij} = \delta_{ij}$

is given by the Kronecker symbol. Since the rotation operators are orthogonal matrices this corresponds to exactly 3 independent conditions. Note that the rotation operations commute with each other. This is so, since the rotation axes are body fixed and not space fixed. A more practical procedure where only 4 conditions must be fulfilled by reducing the number of rotational degrees of freedom from 8 to 6 can be found in Ref. 34.

The Monte Carlo method is performed by scanning with this window algorithm along the protein backbone and shifting the window after each step by one amide plane. The algorithm may be more efficient if the scans are started at both ends of the protein backbone and are performed only up to the center part of the protein sequence. Another option which has not been used in the present application is to scan along the backbone by choosing the windows at random. The latter method can avoid a possible bias in the procedure. Conformational changes in the two terminal windows containing three amide planes from either end of the protein backbone are not subject to constraints. In this case the angles of all 6 rotational degrees of freedom are chosen arbitrarily. The arbitrary angles are randomly chosen from an interval $[-\varphi,+\varphi]$. The performance of the algorithm depends also on the maximum size of the angle interval for the various cases. The maximum angles φ_{end}, φ_{center}^0 and φ_{center}^1 refer to the 6 angles of rotation in a terminal window, the 2 arbitrary angles in a center window and the 4 angles determined in a center window respectively.

5. EQUILIBRIUM PROPERTIES OF THE PROTEIN BACKBONE MODEL

The Monte Carlo method of dynamics simulation with the window algorithm is by far not the most efficient technique to obtain equilibrium properties of a protein backbone in the absence of non-bonded interactions. The reptation algorithm is for instance more efficient but cannot easily be generalized to heterogeneous polymer models[21]. Nevertheless it is instructive to calculate equilibrium quantities and to check in this way the degree of ergodicity and the statistical accuracy which can be obtained with the window algorithm. If not otherwise stated all calculated quantities are obtained from a statistical ensemble involving 2×10^6 scans of the window algorithm along a protein backbone of 24 amide planes.

The most relevant quantities are the first and second order moments $\langle \vec{R}_n^{(N)} \rangle$ and $\langle (\vec{R}_n^{(N)})^2 \rangle$ of the end–to–end distance distribution. Thereby the end-to-end distance vector of n consecutive amide planes

$$\vec{R}_n^{(N)} = \sum_{j=-n}^{+n} \vec{R}_{[(N + j)/2]} \tag{13}$$

refers to the center part (n amide planes) of a larger protein backbone sequence of N amide planes. The bracket [...] stands for the entire function. The $\vec{R}_j$ are the diagonal vectors of the amide planes, $\vec{R}_j = \vec{A}_j + \vec{B}_j$. For the numerical evaluation of the moments we take $\vec{A}_j^2$ $= 0.5 = \vec{B}_j^2$ such that $\vec{R}_j^2 = 1$. In this special case one can obtain the following exact expression for the second order moments[34]

$$\frac{1}{n} \langle (\vec{R}_n^{(N)})^2 \rangle = \frac{1}{4}(6 + \sqrt{3}) - \frac{3}{8n}(2 + \sqrt{3})(1 - 3^{-n}). \tag{14}$$

The exact values of the first order moments vanish.

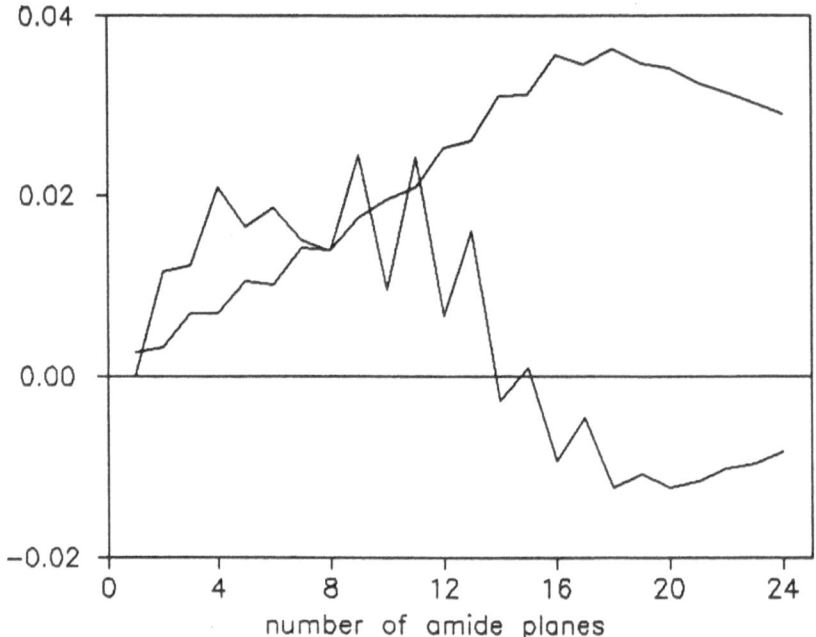

number of amide planes

Fig. 3:   The equilibrium statistics of the end-to-end distance distribution is depicted as a function of the number of amide planes. The considered amide planes are taken from the center part of a simulated protein backbone made of 24 square-shaped amide planes whose diagonal vectors are of unit length. The non-bonded interactions of the model protein vanish. The displayed results are based on a Monte Carlo dynamics of $2\,10^6$ scans of the window algorithm along the protein backbone. The initial configuration is taken randomly as described in the text. The conformations are saved after each 10th scan of the window algorithm. The angle parameters governing the acceptance of conformational changes by the window algorithm are given as $\varphi_{end} = 180^0$, $\varphi_{center}^0 = 15^0$, and $\varphi_{center}^1 = 50^0$. The lower curve represents the deviation of the normalized second moments of the end-to-end distance distribution $\langle (\vec{R}_n^{(N)})^2 \rangle / n$ from the exact values according to Eq. (9). The upper curve represents the normalized scalar product of the first moments $\langle (\vec{R}_n^{(N)}) \rangle^2 / n$.

In Fig. 3 the normalized scalar product of the first order moments $\langle \vec{R}_n^{(N)} \rangle^2 / n$ and the deviations of the calculated normalized second order moments $\langle (\vec{R}_n^{(N)})^2 \rangle / n$ from the exact values, Eq. (14), are depicted as a function of the number n of the considered amide planes. The end-to-end distance vectors of n consecutive amide planes refer to the center part of a larger piece of protein backbone which has been used for the computer simulation.

The deviations of the normalized second order moments are rather small. Their absolute values do not exceed 0.02 and show no systematic n dependence. This tells us that the statistical accuracy of the second order moments depends rather on the size of the total protein backbone unit whose dynamics is simulated. An ensemble of 2000 independent protein backbone conformations obtained with the algorithm described in section 3 provides statistical errors of the same magnitude. Hence the simulated protein backbone conformations are independent of each other after 1000 scans of the window algorithm. This value agrees with the behaviour of the time correlation functions discussed in the next section.

The normalized scalar product of the first order moments exhibits a different behaviour. The statistical error is larger (up to 0.035) and increases roughly linear with the number n of considered amide planes. In the last part from 18 to 24 amide planes the scalar product of the normalized first order moments decreases. This is due to the fact that the motion of the 3 terminal amide planes on both ends of the protein backbone is not subject to any constraints. Hence this part of the backbone relaxes almost instantaneously to equilibrium and cannot contribute to the statistical error.

The statistical error of both the first and the second order moments probe the long time decay and a possible non-ergodic behaviour of conformational fluctuations of the protein backbone model. The first order moments involve also an average over different orientations of the considered part of the protein backbone. They can only vanish if the underlying statistical ensemble covers not only all conformations but also all orientations of the protein backbone. This explains that the statistical errors of the normalized first order moments can be larger than those of the second order moments (see Fig. 3).

6. TIME CORRELATION FUNCTIONS OF PROTEIN DYNAMICS

A time course of conformational changes for a protein backbone of 40 amide planes is depicted in Fig. 4. The initial configuration is a perfectly stretched all trans ($\phi = 180^0 = \psi$) protein backbone conformation. By following the time evolution of the end-to-end distance one observes that the initial length shrinks dramatically during the first few thousand scans of the window. A typical equilibrium value of the end-to-end distance can be obtained by

start configuration   40 planes   141.9 Å

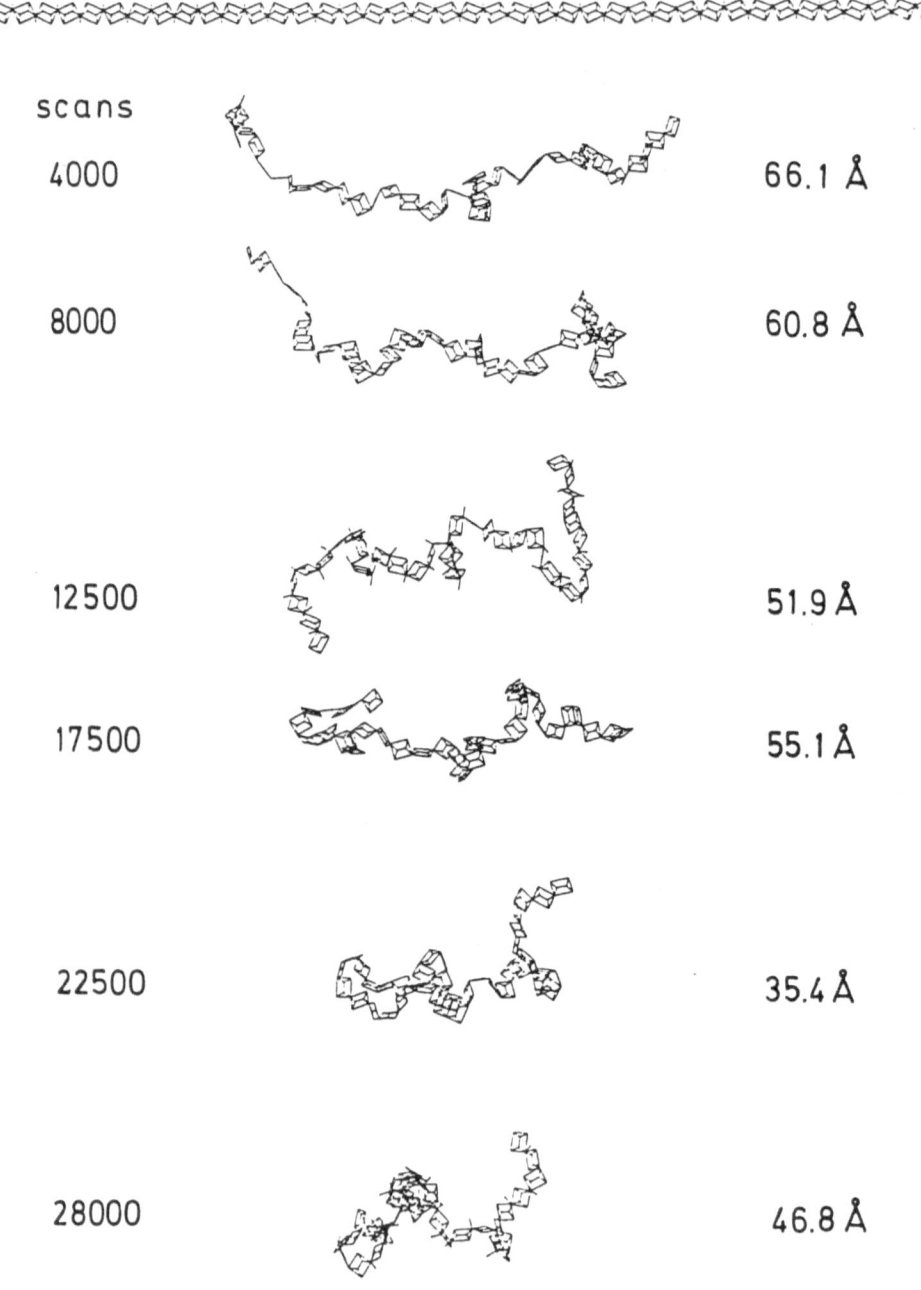

scans

4000   66.1 Å

8000   60.8 Å

12500   51.9 Å

17500   55.1 Å

22500   35.4 Å

28000   46.8 Å

Fig. 4: 7 Snapshots of the time evolution of the all trans ($\phi = 180^0 = \psi$) conformation of 40 amide planes are displayed as a projection on the plane of the initial configuration. The amide planes are of rectangular shape with $\vec{A}_j^{\,2} = (2.8\text{Å})^2$ and $\vec{B}_j^{\,2} = (2.4\text{Å})^2$. The number of scans of the window algorithm along the protein backbone is given at the left side. At the right side the end–to–end distance of the model protein backbone is monitored. The angle parameters governing the window algorithm are $\varphi_{\text{end}} = 60^0$, $\varphi_{\text{center}}^0 = 15^0$, and $\varphi_{\text{center}}^1 = 50^0$. For explanation see text. The probability of acceptance with these parameters is about 0.5.

---

calculating $\langle (\vec{R}_{40}^{(40)})^2 \rangle^{\frac{1}{2}} = 31$ Å analog to Eq. (14). Between 12 500 and 17 500 scans of the window algorithm the end–to–end distance starts to increase for the first time. This marks the onset of equilibrium fluctuations of protein backbone conformations and is related to the second order moments of the end–to–end distance distribution. However, the overall orientation of the configurations is still strongly correlated with the initial configuration, as one can see. The final configuration after 28 000 scans of the window algorithm appears to be more compact than the previous configuration at 22 500 scans, even though the end–to–end distance has opposite values. This indicates that now orientational relaxation of the protein backbone sets in.

To characterize the time evolution of the Monte Carlo dynamics of the protein backbone model, we consider two different autocorrelation functions of the end-to-end distance vectors, Eq. (13). The time correlation functions are obtained by replacing the ensemble average by a time average, with a total time interval of $[0, T]$. In the following formulas the superscript N refering to the lenght of the whole chain has been omitted. The angular autocorrelation function

$$\Phi_1(t) = \left\langle \frac{\vec{R}_n(t) \cdot \vec{R}_n(0)}{\sqrt{(\vec{R}_n(t))^2 \, (\vec{R}_n(0))^2}} \right\rangle_{[0, T-t]} \tag{15}$$

monitors the orientational relaxation of the considered part of the protein backbone in analogy to the first order moments. The subscript at the angular brakets denotes the time interval available for the time average. The other autocorrelation function refers to the second order moments

$$\Phi_2(t) = \frac{\langle R_n^2(t) \, R_n^2(0) \rangle_{[0, T-t]} - \langle R_n^2 \rangle_{[t, T]} \, \langle R_n^2 \rangle_{[0, T-t]}}{\langle R_n^4 \rangle_{[0, T]} - \langle R_n^2 \rangle_{[0, T]}^2} \, . \tag{16}$$

It monitors structural relaxations of breathing motions of the protein backbone model. Such correlation functions have often been considered in polymer science[26,33,34].

For discrete equidistant time points the time correlation functions in expressions (15) and (16) read

$$\langle f(j\,\Delta t)\, f(0)\rangle_{[0,M-j]} = \frac{1}{M+1-j} \sum_{k=0}^{M-j} f((j+k)\Delta t)\, f(k\,\Delta t) \tag{17}$$

The time increment Δt corresponds to one scan of the window algorithm along the protein backbone and M+1 is the total number of considered conformations. Both time correlation functions are normalized such that $\Phi_i(t=0) = 1$ and $\Phi_i(t=\infty) = 0$.

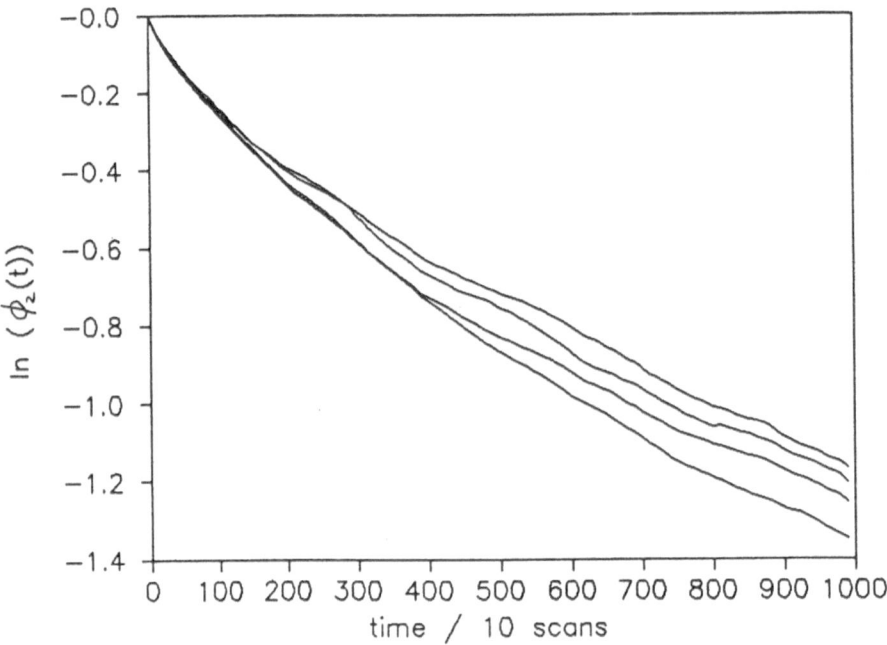

Fig. 5: The time autocorrelation function $\Phi_2(t)$. Eq. (16) of the second moments of the end-to-end distance distribution of 14 amide planes from the center of a protein backbone of 24 amide planes is depicted on a semilogarithmic plot. All parameters governing the Monte Carlo simulation are the same as in Fig. 3. From top to bottom the 4 curves refer to ensemble averages based on Monte Carlo simulation data from the first $1.0\ 10^6$, $0.5\ 10^6$, $1.5\ 10^6$, and all $2.0\ 10^6$ scans of the window algorithm along the protein backbone.

Simulation data corresponding to the elapsed time of the considered time correlation function are no longer available for the time average. Hence depending on the elapsed time one can expect small time dependent drifts in the ensemble averages of the correlation function. They can give rise to unphysical negative values and oscillations of the time correlation function where its value gets close to the statistical error. To avoid these phenomena the mean values of the second order moments which appear in the nominator of the time correlation function (16) are evaluated in a symmetrized form.

In Fig. 5 the time decay of the second order moment correlation function $\Phi_2(t)$ of a chain segment of 14 amide planes located in the center part of a protein backbone of 24 amide planes is depicted. The semilogarithmic plot demonstrates clearly that the decay is non-exponential in time. The same is true for the time decay of $\Phi_1(t)$ (not shown in a figure). By following the build-up of the time correlation function $\Phi_2(t)$ with increasing size of the underlying statistical ensemble one can conclude that the time decay is reliable at least up to 4000 scans. With increasing time there are systematic deviations. Normally the time decay seems to be slower for smaller statistical ensembles.

In Fig. 6 the dependence of $\log_{10}[-\ln(\Phi_2(t))]$ on $\log_{10}(t)$ is depicted. With this type of plot one obtains straight lines for single exponential and stretched exponential decay laws

$$f(t) = \exp[-(t/\tau)^{\alpha}] . \tag{18}$$

The initial time decay pattern up to a value of $\Phi_2(t) = 0.20$ follows very closely a stretched exponential decay law. Below this value the decay slows down. This can be traced back to the quality of the underlying statistical ensemble which is no longer reliable at such small values of the time correlation function. At even smaller values of the time correlation function namely at $\Phi_2(t) = 0.02$ the time decay exhibits a rather irregular behaviour. This is the instance where the time correlation function is merely governed by the statistical error of the second order moments whose deviation from the exact value is just given by $|\langle (\vec{R}_n^{(N)})^2 \rangle_{exp} - \langle (\vec{R}_n^{(N)})^2 \rangle_{theor}| = 0.02$ (see Fig. 3). Without the special second order moments used in the time correlation function, Eq. (16), the value of $\Phi_2(t)$ can become negative and oscillate at longer times. There is also no significant n dependence of the value of the correlation function where the irregular behaviour sets in. This is in agreement with the behaviour of the statistical errors of the second order moments. It confirms our earlier conclusion that with respect to the second order moments and a given length of the simulation the quality of the statistical ensemble depends only on the total length (N = 24) of the simulated protein backbone and not on the lenght of the smaller center part which is under consideration.

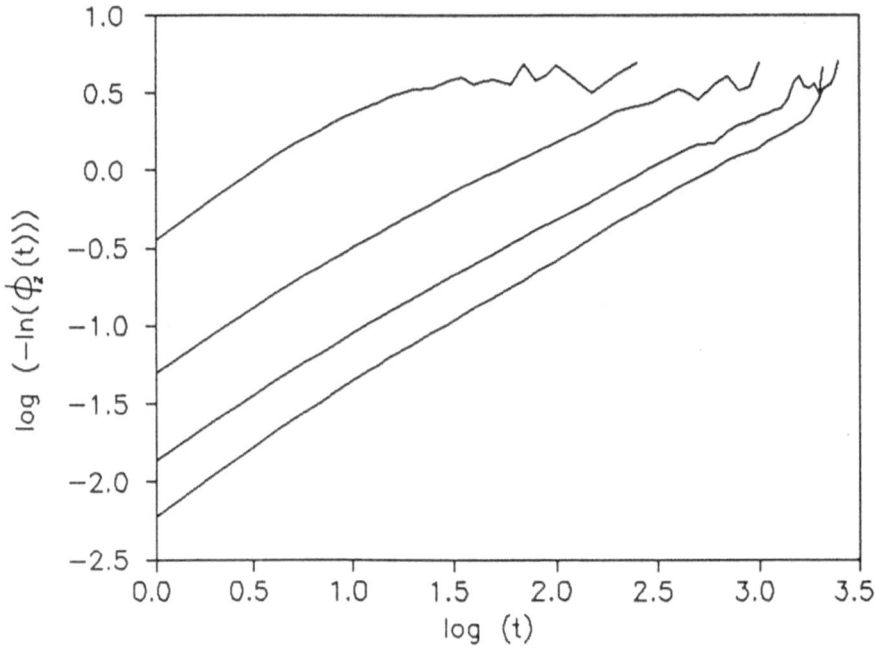

Fig. 6: The time autocorrelation function $\Phi_2(t)$, Eq. (16), is depicted as a function of time. The unit time is here 10 scans of the window algorithm along the protein backbone. The double logarithmic plot is such that a stretched exponential decay appears as a straight line. The parameters governing the Monte Carlo simulation are the same as in Fig. 3. From top to bottom the decay functions refers to a chain length of $2, 4, 8, 16$ amide planes in the center of the simulated protein backbone of 24 amide planes.

The time correlation functions of the first order moments are also plotted such that a stretched exponential decay law appears as a straight line (Fig. 7). In this case the validity of the decay law is not so obvious. In particular very small chain segments of only 1 or 2 amide planes seem to reorient initially relatively fast. However at later times they are slowed down so much by the influence of the other 12 or 11 amide planes which are attached on either side of the considered 1 or 2 amide planes that the character of the decay law changes. This influence diminishes if larger portions of the simulated protein backbone are considered.

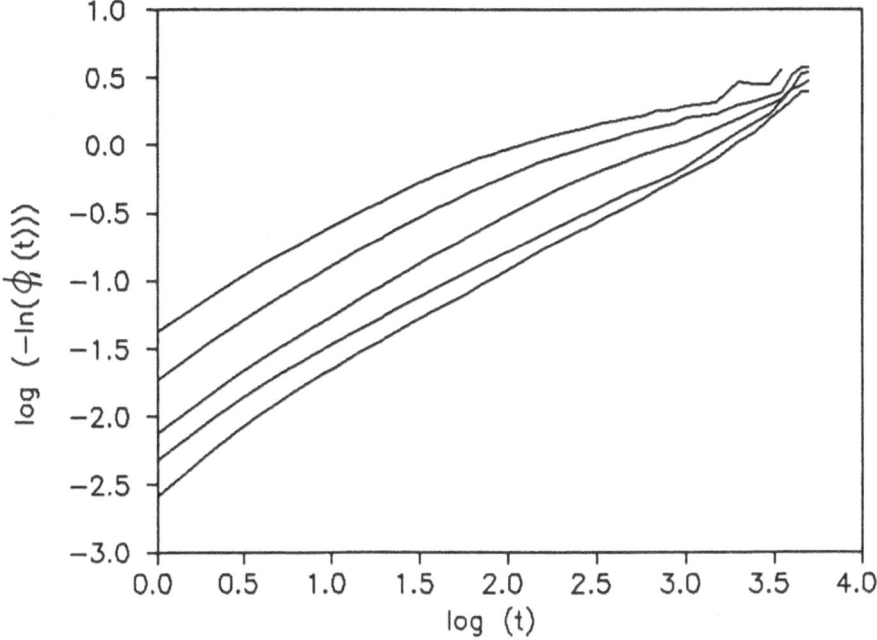

Fig. 7: The time autocorrelation function $\Phi_1(t)$ is depicted in the same way as in Fig. 6. Here the decay functions refer to n = 1, 2, 4, 8, 16 from top to bottom.

By comparing the time decay functions with the stretched exponential decay law, Eq. (18), one obtains values of the exponent α between 0.7 to 0.8. The exponent α exhibits only a very weak, non-systematic n dependence. The α values of the different chain lengths n are very similar for Φ_1 and Φ_2.

It is interesting to speculate on the physical origin of the stretched exponential decay law. Since the exponent α does not depend on the chain length the non-exponential decay law will also be valid for the whole simulated protein backbone consisting of 24 amide planes. The original Rouse polymer model[35] consisting of N beads sequentially connected by harmonic springs without further constraints exhibits a long time decay which is mono-exponential[36]. A kink jump motion of a polymer model on a regular lattice[35] exhibits also a mono-exponential decay at long times. The use of an off-lattice simulation method in the present work should have no influence on this effect. Hence the only influences which remain are the angular constraints of the protein backbone model. They account for interactions induced by the chemical bonding and provide a certain rigidity to the protein backbone model. Due to this rigidity specific relaxation processes in the inner part of the protein backbone may only be possible if certain defects enter or leave the protein backbone at the terminal amide planes. Since the considered protein backbone is of finite length there is in principal a maximum relaxation time beyond which the time decay becomes mono-exponential. However for the present data set the statistical errors are too large to observe this asymptotic behaviour.

To understand the transient time decay of protein backbone conformations a simple model from dielectric relaxation theory can be of use[38]. In this model the reorientation of small molecular groups is only possible in the presence of specific defects. If the defect motion obeys a diffusion process in a one–dimensional space one obtains a stretched exponential decay law for the molecular reorientation with an exponent of $\alpha = 1/2$, Eq. (18). This mechanism may also be the basis for the observed decay law in the present case. By including the non–bonded interactions additional constraints are introduced which will lead to smaller values of α and larger values of τ. To confirm this explanation more intensive computer simulations are necessary.

7. CONCLUSION AND OUTLOOK

The Monte Carlo method with the window algorithm has not been developed to evaluate thermodynamic averages. For polymer and protein models which are as idealized and simple as the present model there are much more efficient methods to accomplish this task[21-30,35,36]. Nevertheless, the thermodynamic averages obtained for short chains are very close to the exact values (see section 5) and demonstrate the ergodic behaviour of the method. Furthermore the method is capable to model a realistic time evolution. It is also interesting to note that even in the absence of non–bonded interactions the constraints accounting for the chemical composition of the protein backbone model give rise to a nontrivial stretched exponential decay pattern.

The window algorithm can also be used for more realistic protein backbone models. For such models the vectors $\vec{a}_j$ and $\vec{b}_j$ spanning the jth amide plane and its vector product $\vec{a}_j \times \vec{b}_j$ serve as a body fixed basis system. This basis system is used to expand the coordinates of the corresponding backbone and residue atoms. Thereby the $C_j(\alpha)$ atoms are situated in the origin of the jth basis system. The backbone atoms C_j, O_j, N_{j+1}, H_{j+1}, and $C_j(\alpha)$ are placed in the jth amide plane. The atoms of the jth residue are placed off plane by using suitable relative vectors in the corresponding body fixed basis system. The atoms are placed after the algorithm for one window has been completed. Then the non–bonded energy terms of the conformational change can be calculated. The new conformation will be accepted according to the Metropolis algorithm[39]. This procedure constitutes a complete Monte Carlo step. In the present simulations where the non–bonded interactions are turned off the probability for acceptance of a Monte Carlo step is typically at 50 %. The energy criterion of the Metropolis algorithm will reduce the acceptance probability considerably. To enhance the acceptance probability one can for instance minimize the conformation energy of the 4 residues which are in the actual window of the protein backbone. This is a mixed type of algorithm. The protein backbone conformations are at non-vanishing temperatures but the residue conformations are locally at zero temperature. Such mixed algorithms have been applied before to study the protein folding problem[5].

In the absence of non–bonded interactions the CPU time for the window algorithm is mainly used to perform the rotation operations in the 3–dimensional space and to solve the constraining conditions. The latter involves the following scheme. The constraining conditions are first linearized. Then the resulting linear equations are solved iteratively. The low dimensions of the rotations and of the linear equations prevent a vectorization of the program code. If the non–bonded interactions are turned on the CPU time is mainly used to evaluate the energy of the protein conformations. As is typical for disordered systems not more than a factor of 4 can be gained by vectorizing the program code. Furthermore one cannot perform several elemantary Monte Carlo steps simultaneously (i.e. in parallel) even if the corresponding windows do not overlap. This is due to the long range character of the non–bonded interactions. These interactions prevent the commutation of Monte Carlo steps of different windows. Hence a naive parallelization of program code introduces a bias.

In spite of these problems parallelization can be of use if the probability of acceptance of a Monte Carlo step is very low. This will be the case in the presence of non-bonded interactions. To overcome the problem with the low acceptance probability one can try several windows in parallel. If more then one window provides a succesful move one must discard the solution from all but one window. Let us assume the probability of acceptance in a single window is on the average p_0 and there are n parallel processes. Then the probability that only a single window provides a succesful move is given by

$$p_1 = \begin{bmatrix} n \\ 1 \end{bmatrix} (1 - p_0)^{n-1} \, p_0 \tag{19}$$

The probability that more than a single window leads to a solution is a measure for the overhead or inefficiency of the method. It is given by

$$P(j>1) = 1 - (1 - p_0)^n - p_1 \, . \tag{20}$$

As an example we take $p_0 = 0.01$ and $n = 10$, then $p_1 = 0.091$ and $P(j > 1) = 0.086$. Hence the theoretical speed up is about a factor of 9.

Another major problem is to find the proper time scale for the Monte Carlo steps. Two strategies have been mentioned in the introduction. One possibility is a comparison with bench mark calculations. The other possibility is a comparison with key experiments. The latter possibility has the advantage to provide a more realistic estimate. On the other hand the actual values for the normation of the time scale depend probably very specifically on the considered protein molecule and cannot be transferred easily to other systems. Furthermore the comparison of such computer simulations and experimental data is limited to certain key features.

With the availability of large amounts of computer time a bench mark calculation based on the propagation of Newton's equations of motion can be performed easily. For this purpose one can in principle also turn off the non—bonded interactions. This reduces the required amount of computer time considerably and the results should be comparable to Monte Carlo simulation data, where only the torsion potentials need to be turned on. But the two different computational methods can only match if the Newtonian dynamics in a window of 3 amide planes is diffusion—like. In the absence of non-bonded interactions the influence of the kinetic energy terms increases. Hence the velocity correlations are very strong and the dynamics is probably no longer diffusion—like. The estimate of the time scale for a Monte Carlo step can then deviate by several orders of magnitude. Nevertheless this method may provide a useful lower limit to estimate the time unit.

Another less expensive variant to gauge the time unit would be to start with a bench mark calculation of a small system and to match the results with a corresponding Monte Carlo calculation. Then the results of the small system are extrapolated to larger systems by using a scaling procedure, based on Stokes law[40]. However, this method can fail if heterogeneous effects from the protein surface become important.

There may even be some time scale problems with the standard method of computer simulation of protein dynamics. Protein dynamics inferred from flourescence depolarisation data in the $10 - 100$ ps time regime can be a factor of 5 slower than data from computer simulation of protein dynamics[41]. At longer time scales the discrepancies could even be larger[33]. A more detailed interaction potential like time dependent atomic partial charges and atomic polarisations may improve on this situation. All of the preceding arguments indicate that the time scale problems can only be solved by a combined effort from several methods including information from experimental data. Nevertheless, we hope to be able to tackle these problems down in the near future.

Acknowledgements

The authors like to thank Prof. S. F. Fischer for continuous support and valuable discussions. The programer skills of Mr. J. Rinser are greatfully acknowledged. Dr. U. Führ and Dr. A. Wadan contributed at an earlier state of this work. The work is financially supported by the Deutsche Forschungsgemeinschaft through SFB 143 C3 and a Heisenberg fellowship.

REFERENCES

1. G. E. Schulz and R. H. Schirmer, Principles of Protein Structure (Springer, Berlin, 1978)
2. R. E. Dickerson and J. Geis, The Structure and Action of Proteins (Harper, New York, 1969)
3. A. Fersht, Enzyme Structure and Mechanism (Freeman, New York, 1984)
4. C. Ghelis and J. Yon, Protein Folding (Academic, New York, 1982)
5. H. A. Scheraga, in: Biological and Artificial Intelligence Systems, E. Clementi and S. Chin. (eds), (Escom, Leiden, 1988) p.1
6. N. Go, Ann. Rev. Biophys. Bioeng. $\underline{1}$, 183 (1983)
7. J. A. McCammon and S. C. Harvey, Dynamics of proteins and nucleic acids (Cambridge, Unipress, 1987)
8. C. L. Brooks III, M. Karplus and B. M. Pettit, Proteins: a Theoretical Perspective of Dynamics, Structure and Thermodynamics (Wiley, New York, 1988)
9. W. F. van Gunsteren, Protein Engineering, $\underline{2}$, 5 (1988)
10. A. Warshel and S. Creighton, in: Computer Simulation of Biomolecular Systems, W. F. van Gunsteren and P. K. Weiner, Eds., (Escom, Leiden, 1989), p.120
11. M. Levitt, Ann. Rev. Biophys. Bioeng. $\underline{11}$, 251 (1982)
12. J. A. McCammon, Rep. Prog. Phys. $\underline{47}$, 1 (1984)
13. S. J. Weiner, P. Kollmann, D. Case, U. Singh, C. Ghio, G. Alagona, S. Profeta, and P. Weiner, J. Am. Chem. Soc. $\underline{106}$, 765 (1984)
14. H. Frauenfelder, F. Parak and R. D. Young, Am. Rev. Biophys. Chem. $\underline{17}$, 451 (1988)
15. A. Ansari, J. Berendzen, S. F. Bowne, H. Frauenfelder, I.E.T. Iben, T. B. Sauke, E. Shyamsunder and R. D. Young, Proc. Natl. Acad. Sci, USA $\underline{82}$, 5000 (1985)
16. M. Fixman, J. Chem. Phys. $\underline{69}$, 1527 (1978)
17. E. Helfand, J. Chem. Phys. $\underline{71}$, 5000 (1979)
18. J. P. Ryckaert, G. Ciccotti, and H. J. C. Berendsen, J. Comp. Phys. $\underline{23}$, 327 (1977)
19. W. F. van Gunsteren and H. J. C. Berendsen, Molec. Phys. $\underline{34}$, 1311 (1977)
20. W. F. van Gunsteren, Molec. Phys. $\underline{40}$, 1015 (1980)
21. A. Kolinski, J. Skolnick, and R. Yaris, Biopolymers $\underline{26}$, 937 (1987)

22. J. Skolnick, A. Kolinski and R. Yaris, Proc. Natl. Acad. Sci. US **85**, 5057 (1988)

23. S. H. Northrup, J. A. McCammon, Biopolymers **19**, 1001 (1990)

24. G. G. Lowry, Ed., Markow Chains and Monte Carlo Calculations in Polymer Science (Dekker, New York, 1970)

25. K. Kremer and K. Binder, Comp.Phys. Rev. **7**, 259 (1988)

26. A. Baumgärtner in: Applications of the Monte Carlo Method in Statistical Physics, K. Binder, Ed. (Springer, Berlin, 1987)

27. A. Baumgärtner, Ann. Rev. Phys. Chem. **35**, 419 (1984)

28. P. G. De Gennes, J. Chem. Phys. **55**, 572 (1971)

29. F. T. Wall and F. Mandel, J. Chem. **63**, 4592 (1971)

30. K. Kremer, A. Baumgärtner and K. Binder, J. Phys. A: Math. Gen. **15**, 2879 (1981)

31. N. Go and H. A. Scheraga, Macromolecules **3**, 178 (1970)

32. W. Braun, Biopolymers **26**, 1691 (1987)

33. F. Parak and E. W. Knapp, Proc. Natl. Acad. Sci. US **81**, 7088 (1984)

34. E. W. Knapp and A. Irgens—Defregger, (1991) in press

35. P. Verdier, J. Chem. Phys. **45**, 2118 (1966); **45**, 2122 (1966); **59**, 6119 (1973)

36. H. J. Hilhorst and J. M. Deutch, J. Chem. Phys. **63**, 5153 (1975)

37. P. E. Rouse, J. Chem. Phys. **21**, 1272 (1953)

38. P. Bordewijk, Chem. Phys. Letters **32**, 592 (1975)

39. N. Metropolis, A. W. Rosenbluth, M. N. Rosenbluth, A. H. Teller, E. Teller, J. Chem. Phys. **21**, 1087 (1953)

40. W. Nadler, A. T. Brünger, K. Schulten, and M. Karplus, Proc. Natl. Acad. Sci. US **84**, 7933 (1987)

41. A. D. MacKerell Jr., L. Nilsson, R. Rigler U. Heinemann, W. Saenger, Proteins **6**, 20 (1989)

Quantum Mechanical Calculations of Small Molecules

G. Frenking
Fachbereich Chemie, Universität Marburg, Hans-Meerwein-Straße,
W-3550 Marburg, FRG

Abstract: The importance of quantum mechanical calculations as a predictive and analytical tool for chemical research is demonstrated by discussing several research projects of our group.

1. INTRODUCTION

The development of quantum mechanical methods has made a stunning progress in the last three decades from an esoteric method which appeared to be useful only for atoms and perhaps hydrogen molecule with little practical use for chemistry to perhaps the most powerful tool for chemical research. While the design of new experiments is usually based on the accumulation of experimental data which are analyzed in terms of simple bonding models, the analysis is much more precise if a proper quantum mechanical calculation is performed. Moreover, the target molecule can be envisaged prior to synthesis and the chance for a successful experiment may be estimated or even predicted. Experiment will never be replaced by theory, but the experimental route can be much better focussed by the help of quantum mechanical studies prior to experiment. The power of modern quantum mechanical methods is based mainly on the development in three different areas:

(i) Progress in hardware architecture, in particular supercomputers and, perhaps more important for chemical research in the 90s, modern workstations.

(ii) Development of user-friendly ab initio (GAUSSIAN [1], CADPAC [2], GAMESS [3]) and semi-empirical (MNDO [4], AM1 [5], PM3 [6]) methods [7].

(iii) Compilation and systematic comparison of theoretical results at various levels of theory with experimental data which allow certain standards in the computational procedures [8].

The dramatic progress in the three areas in particular in the last decade allows the somewhat provocative resume that in 1990 the ground state chemistry of first and second row elements can completely be treated by quantum mechanical methods. By completely we do not mean that molecules of any size can e.g. be geometry optimized at any level of theory. It should be realized that

U. Harms (Ed.)
Supercomputer and Chemistry 2
© Springer-Verlag Berlin Heidelberg 1991

chemical research is mainly concerned with **differences** in energies and properties among different molecules, and that the actual reaction predominantly takes places only in a particular part of the molecule. Truncation of the molecule to the essential part, focussing on differences rather than absolute size of properties in question, and choice of the suitable method are essentials of a skillful computational chemist. While the use of quantum mechanical methods has been made easy and should be a standard research tool for all chemists, a warning should be issued. Behind the apparent simplicity of the programs, there are very complex procedures based upon sophisticated theory. Thus a non-specialist is well advised to consult an experienced theoretician before a conclusion is drawn based on calculated numbers!

2. THEORETICAL MODEL

In the following we want to present a short outline of the essential approximations which are employed on the most common _ab initio_ methods. They are the theoretical basis for all computations which are reported here. It is not the intention to discuss the mathematical background of the procedures, but rather to indicate the basic ideas behind the methods which provide a quantum mechanically based model. A detailed discussion of the quantum mechanical techniques may be found in text books of quantum chemistry [31].

Scheme 1 shows the most important steps in the ladder of approximations to the Schrödinger equation. The first step introduces the independent particle model. In the Hartree-Fock [32] approximation, the many-particle system is reduced to a product of one-particle systems in which one particle, e.g. an electron, interacts with the **average** field of the other electrons. The error introduced by neglecting the individual interactions among all electrons is termed correlation energy, which is the energy difference between the exact (nonrelativistic) value of the energy and the result of the Hartree-Fock calculation of the respective system. There are several methods which allow the calculation of the correlation energy approximately. The most commonly used techniques are based on perturbation theory, in particular Møller-Plesset perturbation theory [36], and configuration interaction (CI) [37]. Further simplifications which are usually introduced at this level of theory are the Born-Oppenheimer [33] approximation, which states that the kinetic energy of the electrons and the nuclei may be calculated separately, and for lighter atoms the neglect of relativistic contributions.

Calculations at the Hartree-Fock level are very impractical for studies of molecules, since the computation of the necessary integro-differential equations is very cumbersome. Most quantum mechanical methods approximate the one-electron functions φ_i ("molecular orbitals") as linear combination of usually atom-centered functions ϕ_μ ("atomic orbitals"; LCAO-method), which reduces the calculation of the approximate wavefunction to the optimization of the expansion coefficients $c_{\mu i}$ in the linear Roothan-Hall [34] equations (Scheme 1). The drawback of the method is the large number of two-electron integrals $<\mu\nu|\lambda\sigma>$ which means that the Roothan-Hall equations are a N^4 problem, N being the number of basis functions. Most <u>ab initio</u> calculations are based on the Hartree-Fock Roothan-Hall method. Thus, the two most important parameters which determine the theoretical level of the quantum chemical method to calculate a molecule with a given geometry are: (i) the method by which the correlation energy is estimated and (ii) the basis set. The knowledge of the accuracy which can be expected from different levels of approximations is an important topics in quantum chemistry [35].

Scheme 1. Hierarchy of approximations in quantum mechanical methods.

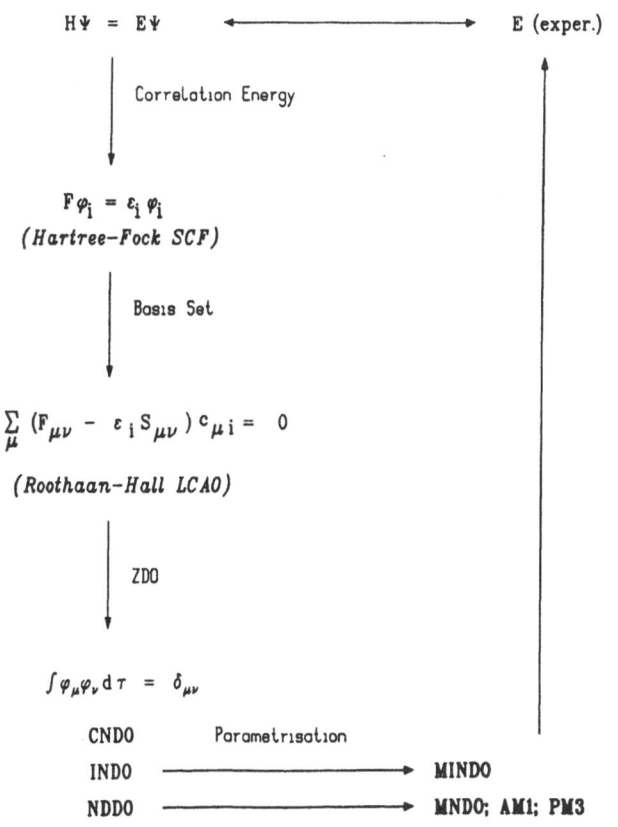

3. Cl_2F^+ - IS IT SYMMETRIC ($Cl-F-Cl^+$) OR ASYMMETRIC ($Cl-Cl-F^+$)?

Although the Cl_2F^+ cation **1** has been prepared in the form of its AsF_6^- and BF_4^- salts already in 1969 [9], it was until recently [10] not clear if the structure corresponds to the C_{2v} form $Cl-F-Cl^+$ **1a** or to the C_s form $Cl-Cl-F^+$ **1b**. While the IR spectra of the salt compounds were interpreted [9] in favor of **1a**, the corresponding Raman spectra indicated [11] that the cation may rather have the structure **1b**. Early calculations also gave conflicting results concerning the relative energies of the two isomers [12]. Thus, the latest edition of a modern textbook in chemistry [29] came to the conclusion that, concerning the symmetric or asymmetric structure of Cl_2F^+, "...the question must be regarded as still open".

We employed modern <u>ab initio</u> techniques to provide a definitive answer to the open question. The results of our calculated geometries and relative energies of **1a** and **1b** are shown in Table 1.

Table 1. Calculated total energies E_{tot} (in Hartrees), relative energies E_{rel} (in kcal/mol), zero-point vibrational energies ZPE (in kcal/mol), bond lengths r_{AB} (in Å), and bond angles $<_{ABC}$ (in degree) for $ClFCl^+$ **1a** and $ClClF^+$ **1b**.

	$Cl-F-Cl^+$ **1a** ($^1A_1$)				
	E_{tot}	E_{rel}	ZPE	r_{ClF}	$<_{ClFCl}$
HF/4-31G[a]	-1016.7509	0.0		1.80	140
HF/DZP[c]	-1017.8336	0.0		1.734	124.2
MP3/DZP[d]	-1018.3122	0.0			
HF/6-31G(d)	-1017.8234	0.0		1.736	122.8
MP2/6-31G(d)	-1018.2782	0.0	2.1	1.767	116.3
HF/6-311G(2df)	-1017.9094	0.0		1.721	123.1
MP2/6-311G(2df)	-1018.6558	0.0		1.745	116.0
MP3/6-311G(2df)	-1018.5393	0.0		1.737	118.6
MP4/6-311G(2df)[b]	-1018.5772	0.0			
MPn/6-311G(2df)[e]	-1018.5818	0.0			

(**Table 1, Cont.**)

$$Cl-Cl-F^+ \text{ 1b } (^1A')$$

	E_{tot}	E_{rel}	ZPE	r_{ClCl}	r_{ClF}	$<_{ClClF}$
HF/4-31G[a]	-1016.7358	9.5		2.270	1.704	99
HF/DZP[c]	-1017.8714	-23.7		1.989	1.565	102.3
MP3/DZP[d]	-1018.3617	-31.1				
HF/6-31G(d)	-1017.8562	-20.6		1.968	1.568	102.3
MP2/6-31G(d)	-1018.3210	-26.9	2.3	1.962	1.614	103.9
HF/6-311G(2df)	-1017.9699	-38.0		1.947	1.537	103.1
MP2/6-311G(2df)	-1018.7335	-48.8		1.924	1.584	104.7
MP3/6-311G(2df)	-1018.6120	-45.6		1.953	1.579	103.9
MP4/6-311G(2df)[b]	-1018.6460	-43.2				
MPn/6-311G(2df)[e]	-1018.6495	-42.5				

[a]Ref. 12a; [b]Using geometries optimized at MP3/6-311G(2df); [c]Ref. 12e; [d]Ref 12e, using geometries optimized at HF/DZP; [e]Energy calculated using equation 6.29 of ref. 8 for estimate of total correlation energy.

The predicted stability difference converges rather smoothly and shows that, with inclusion of zero-point corrections ZPE, **1b** is 43 ± 5 kcal/mol lower in energy than **1a**. This is a strong indication that the cation in the corresponding salt compounds has the C_s form $Cl-Cl-F^+$. This conclusion is further corroborated by the comparison of the theoretically predicted and experimentally observed vibrational spectra, i.e. the Infrared and Raman frequencies and intensities which are shown in Table 2.

Table 2. Experimentally observed and theoretically predicted vibrational spectra for Cl_2F^+. Frequencies are given in cm^{-1}, IR-intensities in km/mol, Raman-intensities in $Å^4$/amu.

$$Cl-F-Cl^+ \text{ 1a}$$

Experiment[a]	Calculated[d]		
IR-frequency(intensity)	ν	IR-Int.	Raman-Int.
ν_1: 520(vw)[b]; 527(mw)[b]; 535(m)[b] 519(m)[c]; 528(mw)[c]; 532(mw,sh)[c]	577(a_1)	5.8	11.7
ν_2: 258(mw)[b]; 293(m)[b]	226(a_1)	0.1	3.2

(**Table 2. Cont.**)

ν_3: 586(mw)[b]; 593(m)[b]   684(b$_2$)   99.4   1.5
588(w,sh)[c]; 594(m)[c]

Cl-Cl-F$^+$ **1b**

Experiment[e]		Calculated[d]		
Raman-bands(rel. intensities)		ν	IR-Int.	Raman-Int.
ν_1: 744(78)[b] 743(90)[c]		806(a')	44.8	17.9
ν_2: 528,535(100)[b] 516(100)[c]; 540(90)		534(a')	13.3	18.7
ν_3: 293,299(20)[b] 296(35)[c]		263(a')	6.9	4.1

[a]Data and assignments are taken from ref. 9; [b]Cl$_2$F$^+$AsF$_6^-$; [c]Cl$_2$F$^+$BF$_4^-$; [d]Frequencies and IR-intensities are calculated at MP2/6-31G(d), Raman-intensities at HF/6-31G(d); [e]Data and assignments are taken from ref. 11.

The calculated intensities, but not the frequencies allow a discrimination between the assignments of the two isomers. Within the error bars [13], the experimental and theoretical frequencies agree for **1a** and **1b**. However, the theoretically predicted IR spectra shows a dominating ν_3 mode for **1a**, while ν_2 should hardly be visible (Table 2). The experimentally obtained spectrum for Cl$_2$F$^+$ shows that ν_2 and ν_3 have comparable medium to medium-weak intensity, which is incompatible with the calculated results for **1a**. In contrast, the theoretically predicted intensities of the Raman lines for **1b** agree perfectly with experiment (Table 2). Thus, the Cl$_2$F$^+$ cations in the salt compounds Cl$_2$F$^+$ AsF$_6^-$ and Cl$_2$F$^+$ BF$_4^-$ [9] have asymmetric (C$_s$) Cl-Cl-F$^+$ structures!

1a or 1b?

1a   **1b**

One further advantage of quantum mechanical calculations of molecules lies in the fact that the analysis of the computed data makes it possible to get much information, part of it may be difficult or impossible to obtain by experiment. In the present case, the calculations were used to study the bond strength in **1a** and **1b**. By using isogyric [14] and isodesmic [15] reactions, the Cl-F and Cl-Cl bond strength can be estimated rather reliably. For **1a**, isodesmic reaction (1) is used:

$$Cl_2F^+ \ \textbf{1a} \ + F(^2P) \ \text{-->} \ ClF^+(^2\Pi) + ClF(^1\Sigma^+_g) \quad -39.1 \ kcal/mol \quad (1)$$

Combination of the calculated energy for reaction (1), $\Delta E_0 = -39.1$ kcal/mol, and the experimentally observed dissociation energy of $ClF(^1\Sigma^+_g)$, $D_0 = 60.3$ kcal/mol [16], gives a theoretically predicted Cl-F bond energy of 21.2 kcal/mol for **1a**.

For **1b**, reactions (2) and (3) are employed to predict the Cl-F and Cl-Cl bond dissociation energy, respectively:

$$Cl_2F^+ \ \textbf{1b} \ + Cl(^2P) \ \text{-->} \ Cl_2^+(^2\Pi_g) + ClF(^1\Sigma^+_g) \quad -18.7 \ kcal/mol \quad (2)$$
$$Cl_2F^+ \ \textbf{1b} \ + Cl(^2P) \ \text{-->} \ ClF^+(^2\Pi) \ + Cl_2(^1\Sigma^+_g) \quad +6.2 \ kcal/mol \quad (3)$$

The combination of the calculated reaction energy $\Delta E_0 = -18.7$ kcal/mol for reaction (2) and the observed bond strength of ClF ($D_0 = 60.3$ kcal/mol) [16] gives a theoretically predicted Cl-F bond strength of 41.6 kcal/mol for **1b**. From the reaction energy $\Delta E_0 = +6.2$ kcal/mol for reaction (3) and the dissociation energy of Cl_2 ($D_0 = 57.1$ kcal/mol) [16] a bond energy of 63.3 kcal/mol is predicted for the Cl-Cl bond in **1b**.

The Cl-F bond in **1a** is very weak (21.2 kcal/mol), while it is twice as strong in **1b** (41.6 kcal/mol). Yet, the Cl-F bond in **1b** is still weaker than in neutral ClF (59.4 kcal/mol [16]) or cationic ClF^+ (68.4 kcal/mol [16]). The Cl-Cl bond in **1b** is stronger (63.3 kcal/mol) than the Cl-F bond. It is also slightly stronger than the Cl-Cl bond in Cl_2 (58.0 kcal/mol [16]), but significantly weaker than the Cl-Cl bond in Cl_2^+ (90.1 kcal/mol [16]).

4. THE STRUCTURE OF DICHLORONITRONIUM ION ("INORGANIC PHOSGENE") Cl_2NO^+

In 1977, Dehnicke et al. [17] reported the first and until now only synthesis of a salt compound containing the dichloronitronium ion Cl_2NO^+ **2**. The X-ray structure of $Cl_2NO^+SbCl_6^-$ shows a unit cell with two symmetry-independent sorts of cations, the small differences between the two structures of Cl_2NO^+ lie within the range of the standard deviations. The main features of the reported geometries are shown in Table 3, together with our [18] calculated data at the MP3/6-31G(d) level, which we consider as reliable.

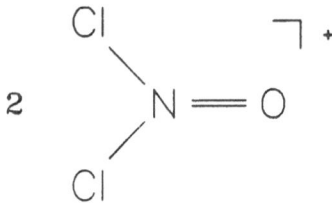

Table 3. Experimentally observed[a] and theoretically predicted[b] bond length [Å] and bond angles [°] of the Cl_2NO^+ cation **2**.

variable	type I[a]	type II[a]	calculated[b]
N-O	1.30	1.32	1.168
N-Cl[1]	1.61	1.64	1.705
N-Cl[2]	1.72	1.69	1.705
O-N-Cl[1]	124	124	122.0
O-N-Cl[2]	118	119	122.0

[a]ref. 17. [b]at MP3/6-31G(d), ref. 18.

The crystal structure shows that the geometry of **2** is distorted from C_{2v} symmetry with one short N-Cl bond length (1.61 - 1.64 Å) and one long N-Cl distance (1.69 - 1.72 Å). The molecule is nearly planar [17]. The calculated geometry for **2**, which is predicted to have C_{2v} symmetry, agrees with the reported data with one notable exception. The N-O distance is computed much shorter (1.168 Å) than it was observed experimentally (1.30 - 1.32 Å). While the theoretically obtained value of 1.168 Å suggests that **2** contains a NO double bond (N-O distance in NO gas 1.151 Å [16]), the experimental value lies between the length of a NO double bond and that of a NO single bond (H_2N-OH 1.47 Å [19]). Thus, theory and experiment indicate qualitatively different bonding in **2**. What is the reason for the discrepancy?

In order to test the reliability of the theoretically predicted N-O bond distance we calculated different types of molecules at various levels of theory which contain the NO moiety. In particular, we optimized the geometries of XNO, X_2NO^+ and X_3NO (X = H, F, Cl) at HF/6-31G(d), MP2/6-31G(d), and MP3/6-31G(d) and compared the results with experimental values. The theoretically predicted and experimentally obtained data for the N-O distances are shown in Table 4.

Table 4. Calculated[a] and experimentally observed N-O bond lengths [Å].

Molec.	HF/6-31G(d)	MP2/6-31G(d)	MP3/6-31G(d)	Exp.
HNO	1.175	1.237	1.213	1.212[b]
FNO	1.128	1.159	1.153	
ClNO	1.16	1.15	1.145	1.14[c]
H_2NO^+	1.149	1.202	1.184	
F_2NO^+	1.110	1.154		
Cl_2NO^+	1.135	1.194	1.168	1.30-1.32[d]
H_3NO	1.376	1.361	1.377	
F_3NO	1.171	1.164	1.168	1.16[e]
Cl_3NO	1.189	1.180		

Table 5. Calculated[a] and experimentally observed vibrational N-O stretching frequencies for Cl_2NO^+ and F_3NO.

	HF/6-31G(d)	MP2/6-31G(d)	Exper.
Cl_2NO^+	2037	1506	1650[d]
F_2NO^+	2248	1800	1862[e]
F_3NO	1664	1914	1690[f]

[a]ref. 18. [b]ref. 30. [c]ref. 31. [d]ref. 17. [e]ref 48. [f]ref. 20.

The most important conclusion which can be drawn from the results shown in Table 4 is that the MP3/6-31G(d) values for the N-O distance are in all cases where experimental values are available very close to the observed data, except for **2**. Only for Cl_2NO^+ there is a large discrepancy between the computed and experimentally observed N-O distance. The N-O distance calculated at MP3/6-31G(d) lies nearly always between the HF/6-31G(d) and MP2/6-31G(d) values. Therefore, we consider the MP3/6-31G(d) data as the most reliable.

In order to test the importance of further extending the basis set, in particular increasing the number of polarization functions, we optimized **2** at HF/6-31G(2df) [18]. In the next paragraph, the reader will find an example where the addition of polarization functions may have a dramatic influence on the computed results. However, the optimized N-O distance in **2** at HF/6-31G(2df) is even shorter (1.126 Å) than at HF/6-31G(d)

(1.135 Å). Thus, the discrepancy between theory and experiment becomes even larger when extending the basis set.

Another source for the difference in the N-O bond length may be that packing forces in the crystal enforce a stretching of the NO bond. The importance of packing forces is obvious by the reported deviation of 2 from C_{2v} symmetry in the crystal [17]. Since one N-Cl bond is significantly longer (1.69 - 1.72 Å) than the other (1.61 - 1.64 Å), we optimized the geometry of 2 with one N-Cl bond frozen either with a short (1.60 Å) or long (1.75 Å) interatomic distance. Calculations at HF/6-31G(2df) and MP2/6-31G(d) both gave the N-O distance and the optimized N-Cl bond with very small differences compared to the completely optimized geometry. As the next step, we studied theoretically the influence of long distance contacts between 2 and counter anions by placing HCl molecules with appropriate Cl-Cl distances around 2 and optimized the geometries at HF/6-31G(d). The unit cell [30] shows that the Cl atom of the long N-Cl bond of 2 is an average of 3.33 Å away from one of the chlorine atoms of $SbCl_6^-$. Furthermore, there are three other Cl atoms of $SbCl_6^-$ anions with a distance of less than 4.0 Å away from either the nitrogen or oxygen atom of 2 [20]. The effect on the computed N-O distance was negligible, the differences were less than 0.02 Å relative to the C_{2v} optimized value [18].

At this point we started to doubt the experimentally obtained [17] N-O distance reported for $Cl_2NO^+SbCl_6^-$. X-ray structure determination is usually a very reliable method to obtain molecular geometries, but errors are possible. In the experimental paper by Dehnicke et al. [17], the authors also report the measured vibrational frequency for the NO stretching mode of 2. The reported value of 1650 cm^{-1} was compared with the corresponding NO stretching mode in F_3NO which is reported as 1690 cm^{-1} [20]. Because of the comparable magnitude for the NO stretching vibration, the NO bond length may be similar in F_3NO and 2. Table 4a shows that the experimentally observed N-O distance in F_3NO is 1.16 Å [20], in excellent agreement with our theoretically obtained values!

Table 5 lists our computed frequency for the NO stretching vibration in 2 and the fluorine analogue F_2NO^+ in comparison with experiment [48]. The results show that the calculated NO stretching frequencies are very sensitive to the equilibrium NO distance obtained at the corresponding theoretical procedure. At HF/6-31G(d), the frequencies are much higher than the experimental values, while at MP2/6-31G(d) they are too low. The lower N-O frequencies at MP2/6-31G(d) correspond to the calculated bond lengths which are significantly longer than at

HF/6-31G(d). However, the frequency shift in going from **2** to F_2NO^+ obtained experimentally as 212 cm^{-1} towards higher wavenumbers (Table 5) is in agreement with the calculated differences of 211 cm^{-1} (HF/6-31G(d)) and 294 cm^{-1} (MP2/6-31G(d)). The theoretical results indicate that the N-O distance in **2** should be slightly longer than in F_2NO^+, but not as much as 1.30 - 1.32 Å as reported[1].

For F_3NO, the HF/6-31G(d) value of 1664 cm^{-1} is lower than observed, while the MP2/6-31G(d) result is too large (Table 5). This trend is in qualitative agreement with the calculated N-O bond distance of F_3NO, which is shorter at MP2/6-31G(d) than at HF/6-31G(d). The opposite trend is found for **2**. Here, the N-O distance is longer at MP2/6-31G(d) than at the HF/6-31G(d) level. Accordingly, the stretching frequency is too high at HF/6-31G(d) and too low at MP2/6-31G(d). The conclusion is that the correct N-O bond length should be intermediate between the two values, and that the distance calculated at MP3/6-31G(d) (1.168 Å) should be very close to the correct value.

Figure 1. Optimized [18] structure of neutral($^1A'$) Cl_2NO **2N**. Distances in Å, angles in degree. The results are obtained at MP3/6-31G(d), the data at MP2/6-31G(d) and at HF/6-31G(d) (in italics) are given in parentheses.

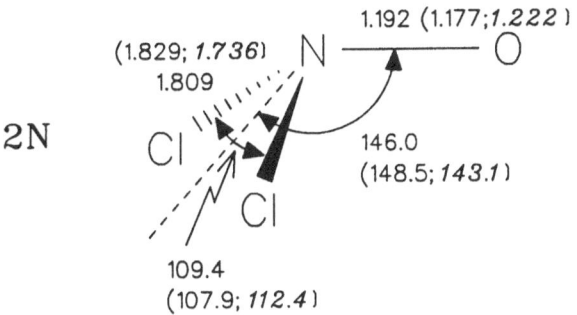

We also calculated the equilibrium geometry and vibrational frequencies of neutral ($^2A'$) Cl_2NO **2N**, which has not yet been studied theoretically. The optimized structure is shown in Figure 1. **2N** has a nonplanar C_s geometry with a N-O distance of 1.192 Å (MP3/6-31G(d)). Again, the MP3/6-31G(d) optimized geometry for **2** predicts values which are intermediate between the data calculated at HF/6-31G(d) and MP2/6-31G(d) (Figure 1). The N-O distance is predicted as 1.222 Å at HF/6-31G(d). The corresponding stretching frequency is calculated as 1366 cm^{-1} (HF/6-31G(d). This means that a N-O bond slightly longer than in **1** should have a stretching frequency which is significantly lower than reported for **1**. The reported N–O distance of 1.30 – 1.32 Å, which is even longer than calculated for **2**, should have a stretching frequency which is still lower, perhaps at 1000 – 1200 cm^{-1}. The experimentally observed value of 1650 cm^{-1} is incompatible with the reported N–O distance, but agrees very well with our calculated bond length.

Based on the comparison of experimentally obtained and theoretically predicted N–O bond lengths in various molecules shown in Table 4, and by the experimental and calculated vibrational frequency for the NO stretching mode in **2**, F_2NO^+, and F_3NO (Table 5), we challenge the reported [17] N–O bond distance in $Cl_2NO^+SbCl_6^-$ obtained by X-ray analysis and predict that the correct value is 1.16 ± 0.03 Å! A reinvestigation of the structure of **2**, perhaps with a different counter ion, could help to resolve the discrepancy.

5. THE RELATIVE STABILITY OF $C_2H_2S_2$ ISOMERS 1,2-DITHIOGLYOXAL AND 1,2-DITHIETE – A NIGHTMARE FOR STANDARD AB INITIO PROCEDURES

The systematic comparison of the performance of different levels of theory for the calculation of molecules containing first and second row elements [8] established certain standards of theoretical levels which enable the theoretician to estimate the accuracy of the computed data. With the knowledge of the reliability of the employed procedure and after carefully check- ing the calculated data, the theoretician may even challenge the results of experiments as described in the previous paragraph [18]. However, blind trust into standard levels of theory may produce ill-devised predictions. The following example may serve as a warning against indiscriminate use of standard <u>ab initio</u> procedures. The example concerns the relative energy of two isomers of the formula $C_2H_2S_2$, i.e. trans 1,2-dithioglyoxal **3a** and the cyclic 1,2-dithiete **3b**.

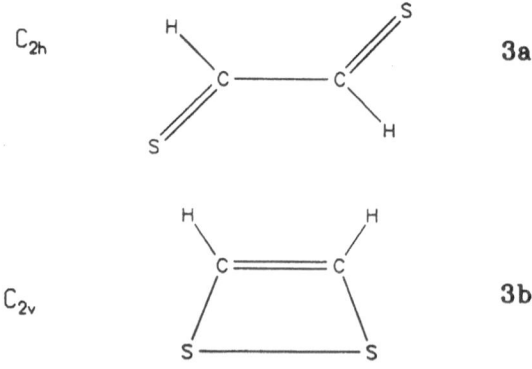

There is compelling experimental evidence provided by photoelectron (PE) spectroscopy [21], gas phase microwave studies [22] and by low-temperature matrix IR spectroscopy [23a] that **3b** is more stable than **3a**. A tentative interpretation of the IR spectrum of $C_2H_2S_2$ produced by pyrolysis of 1,3-dithiol-2-one showed that 90% of the probe consists of **3b** and perhaps 10% of **3a**, but the bands assigned to **3a** could also be caused by unidentified products [23a].

Table 6. Calculated[a] relative energies E_{rel} (kcal/mol) of structures **3a** and **3b**.

	3a	**3b**
	E_{rel}	E_{rel}
HF/3-21G//HF/3-21G[b]	0.0	6.6
HF/6-31G(d)//HF/6-31G(d)[b]	0.0	5.0
MP2/6-31G(d)//MP2/6-31G(d)[c]	0.0	5.5
MP4(SDQ)/6-31G(d)//HF/6-31G(d)[b]	0.0	6.1
HF/6-31G(d)//MP2/6-31G(d)	0.0	5.00
HF/6-31G(2d)//MP2/6-31G(d)	0.0	2.30
HF/6-31G(3d)//MP2/6-31G(d)	0.0	3.48
HF/6-31G(2df)//MP2/6-31G(d)	0.0	-0.91
HF/6-31G(3df)//MP2/6-31G(d)	0.0	-0.61
HF/6-31G(2d2f)//MP2/6-31G(d)	0.0	-1.40
MP2/6-31G(d)//MP2/6-31G(d)	0.0	5.46
MP2/6-31G(2d)//MP2/6-31G(d)	0.0	0.71
MP2/6-31G(3d)//MP2/6-31G(d)	0.0	3.40
MP2/6-31G(2df)//MP2/6-31G(d)	0.0	-6.80
MP2/6-31G(3df)//MP2/6-31G(d)	0.0	-5.36
MP2/6-31G(2d2f)//MP2/6-31G(d)	0.0	-8.46

[a]Unless otherwise noted, data are taken from ref. 27. [b]ref. 25. [c]ref. 23b

Standard _ab initio_ methods predict a different stability order (see the first four entries in Table 6) [24]. At HF/3-21G(d)//HF/3-21G(d), **3a** is calculated as 6.6 kcal/mol lower in energy than **3b** [25]. This result changes little when the basis set is improved to HF/6-31G(d)//HF/6-31G(d) (ΔE = 5.0 kcal/mol in favor of **3a** [25]) and when correlation energy is included. At MP2/6-31G(d)//MP2/6-31G(d), **3a** is 5.5 kcal/mol more stable than **3b** [6], and at MP4(SDQ)/6-31G(d)//HF/6-31G(d), **3b** is higher by 6.1 kcal/mol [25]. The inclusion of zero-point vibrational energies leads to an even higher stability of **3b** by 0.7 kcal/mol (3-21G(d)) [25]. Problematic is not only the wrong prediction that **3a** is considerably more stable than **3b**, but the fact that the theoretical results seem to have converged. Although errors of up to 4 kcal/mol at MP4(SDQ)/6-31G(d) have been reported for the relative energies of different isomers [26], the stability order is usually predicted in the right way. Without the experimental results, one would probably consider the theoretical results as sufficiently reliable to predict that **3a** is more stable than **3b**.

The puzzling failure of the quantum mechanical methods prompted us to start a more detailed investigation. Since neither the extension of the basis set nor inclusion of correlation energy at higher order perturbation theory did change the result significantly, the effect of adding further polarization functions was studied [27]. At the HF/6-31G(d) level, **3a** is predicted as 5.00 kcal/mol more stable than **3b**. The addition of a second d-type function has a sizable influence on the data. **3a** is only 2.30 kcal/mol lower in energy than **3b** at HF/6-31G(2d), and 0.71 kcal/mol at MP2/6-31G(2d). This trend is reversed when a third d-type function is added. **3a** is more stable than **3b** by 3.48 kcal/mol (HF/6-31G(3d)) and 3.40 kcal/mol (MP2/6-31G(3d)). The most dramatic effect on the computed energy difference between **3a** and **3b** comes from the addition of a set of f-functions. The cyclic isomer **3b** is predicted as more stable than **3a** if a 6-31G(2df) or 6-31G(3df) basis set is used. At the Hartree-Fock level, **3b** is favored by 0.91 kcal/mol (HF/6-31G(2df)) and 0.61 kcal/mol (HF/6-31G(3df)). With inclusion of correlation energy at the MP2 level, **3b** becomes <u>much</u> lower in energy than **3a**. The energy difference in favor of **3b** is 6.80 kcal/mol at MP2/6-31G(2df), and at MP2/6-31G(3df) it is 5.36 kcal/mol. The addition of a single set of f-type polarization functions changes the relative energies of **3a** and **3b** by 7.51 kcal/mol at MP2/6-31G(2df) and even 8.76 kcal/mol at MP2/6-31G(3df)!

The effect of adding a second set of f-functions on the basis set was also studied [27]. At HF/6-31G(2d2f), **3b** is 1.40 kcal/mol lower in energy than **3a**, while at MP2/6-31G(2d2f), the energy difference in favor of **3b** becomes 8.46 kcal/mol (Table 6). Thus, a second set of f-type polarization functions increases the stability difference between **3a** and **3b** further, although the effect is not as strong anymore as for the first set of f-functions. It seems that the series of adding polarization functions has nearly converged with 3 sets of d-functions and two sets of f-functions. With the assumption that the theoretically predicted energy difference of 8.46 kcal/mol at MP2/6-31G(2d2f) would be reduced by ~1.5 kcal/mol if a third set of d-functions would be added, our best estimate is that **3b** is 7.0 kcal/mol more stable than **3a**. This reduces after zero-point energy correction of 0.7 kcal/mol [25] to a theoretically predicted value of 6.3 kcal/mol in favor of **3b**, which is in striking contrast to earlier studies [23b,25] predicting that **3a** is more stable than **3b** by 5.0 - 6.8 kcal/mol. Although the correlation energy has been calculated only at the MP2 level of theory, we do not think that the results change significantly if higher order perturbation theory is employed, or if different methods are used. Previous studies have shown [25] that the calculation of correlation energy by different methods do not change the relative energy of **3a** and **3b** much. The clearly higher stability of **3b** over **3a** predicted by the data shown in Table 6 [27] means that the two bands at 826 and 1223 cm^{-1} observed in the IR spectrum of $C_2H_2S_2$, which have tentatively been assigned to a 10% presence of **3a** [23a], probably belong to unidentified compounds.

We would like to comment on the result that the correct stability order **3b** > **3a** is obtained only when a set of f-functions is added. Does this mean that f-type orbitals are important for the description of the electronic structure of sulfur compounds? The answer is no! The importance of including f-type functions is a mathematical one allowing a better description of the wave function. As shown by Reed and Schleyer [28], even d-orbitals are not very important for a description of bonding in molecules containing third-row elements.

6. THE STRUCTURES OF C_6S_6 AND C_6O_6

The search for binary compounds of carbon with elements other than ubiquitous hydrogen has always fascinated chemists. The question 'how many carbon oxides are stable' belongs to this area of research. The number of carbon-sulfur species is a related topics. Very recently, the synthesis of C_2S_2 has been reported [38]. The experimental success of this elusive species was made particularly thrilling by quantum mechanical calculations which suggest that C_2S_2 has a singlet ground state, which would be the first example of a stable molecule violating Hund's rule [38].

Here we want to show how <u>ab initio</u> calculations could help to answer the question on the structure of another interesting carbon sulfur molecule, i.e. C_6S_6. Very recently, Sülzle et al. [39] reported on the mass spectrometric investigation of benzoetrisdithiocarbonic acid **4**, which, after loss of three CO from the radical anion **4**[-.] and subsequent charge exchange reactions, yields the fragments $C_6S_6^n$ (n = +1, 0, -1). The experimental findings provided no information about the structure of the $C_6S_6^n$ fragments. As possible structures the valence isomers **5a** and **5b** were proposed [39].

Figure 2. Calculated [40] bond length in Å for **5a** - **5e** at HF/3-21G(d) (STO-3G). Relative energies in kcal/mol at MP2/3-21G(d)//HF/3-21G(d) + ZPE.

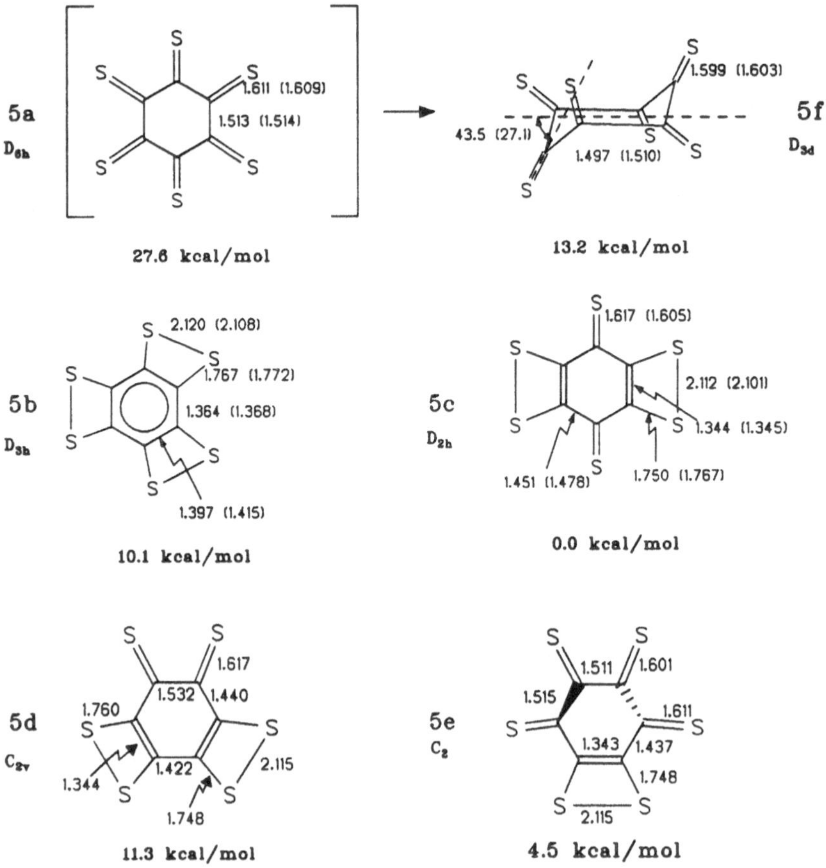

5a and 5b are not only aesthetically attractive, their relative stabilities are also of considerable interest for electronic structure theory. Therefore, we calculated the geometries and vibrational frequencies of the C_6S_6 isomers **5a** and **5b** at the HF/3-21G(d) level of theory, with d-functions only at sulfur [40]. The relative energies of the isomeric forms were then computed at MP2/3-21G(d)//HF/3-21G(d). We also studied theoretically the isomeric forms **5c**, **5d**, and **5e**. In addition, we compared the results for C_6S_6 with valence isoelectronic C_6O_6. The calculated results for C_6S_6 isomers are shown in Figure 2.

The theoretical results may be summarized as following:

(a) The hexathioketo isomer **5a** is not planar (D_{6h}), but has a chair-type geometry **5f** (D_{3d}) similar as in cyclohexane.

(b) The energetically lowest-lying form is <u>neither</u> **5a** nor **5b**, but rather the para-chinoid isomer **5c**.

(c) **5c** is 4.5 kcal/mol more stable than the next isomer **5e**. The energy differences between **5b**, **5d**, **5e**, and **5f** are rather small.

Thus, it may be concluded that the C_6S_6 isomer observed in the mass-spectroscopic experiments by Sülzle et al. [39] has the structure **5c**. Shortly after the results of the theoretical study had been published [40], C_6S_6 was prepared in matrix isolation studies [41] from the same precursor compound **4**. It was possible to measure the infra-red spectrum of the C_6S_6 isomer. In addition, it was found that the isolated C_6S_6 could be excited using UV radiation yielding a second C_6S_6 isomer which is higher in energy [41]. The comparison of the measured IR spectrum with the theoretically obtained vibrational frequencies clearly showed that the most stable form has the predicted low-energy structure **5c**, and that the excited form corresponds to the aromatic structure **5b** [42]. Thus, the information gained from theory and experiment could be combined like pieces of a puzzle, the result being a clear analysis of the observed phenomena.

Could analogous experiments be successful searching for C_6O_6 species? Most probably not! The calculated data [40] in Figure 3 show the computed structures and relative energies of C_6O_6 isomers **6a** - **6c**. The hexaketoform **6a** is clearly the energetically lowest-lying form. However, **6a** is not a very stable species. The dissociation of **6a** into 6 molecules CO is endothermic by 27.5 kcal/mol (MP2/3-21G(d)//HF/3-21G(d)), which suggests that the mean CC bond energy is only $27.5/6 = 4.6$ kcal/mol. In contrast the CC bond energy in **5c** is much higher, $221.8/6 = 37.0$ kcal/mol. Thus, **6a** may not survive under the reaction conditions necessary to abstract 6 CO from the precursor compound.

5c ----->	6CS	+221.8 kcal/mol
6a ----->	6CO	+27.5 kcal/mol

Figure 3. Calculated [40] bond length in Å for **6a** - **6c** at HF/3-21G(d) (STO-3G). Relative energies in kcal/mol at MP2/3-21G(d)//HF/3-21G(d) + ZPE.

6a

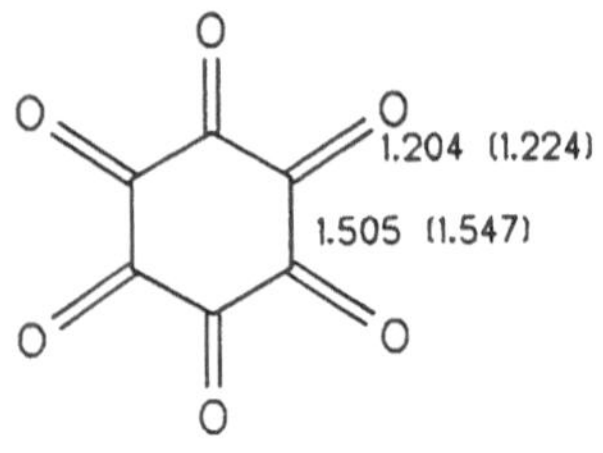

0.0 kcal/mol

6b

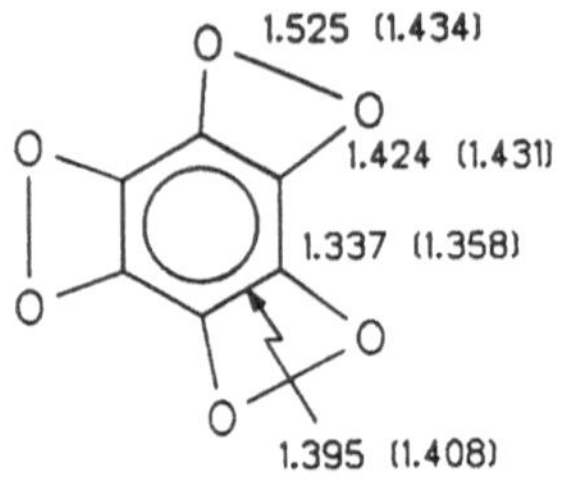

1.525 (1.434)
1.424 (1.431)
1.204 (1.224)
1.505 (1.547)
1.337 (1.358)
1.395 (1.408)

233.9 kcal/mol

6c

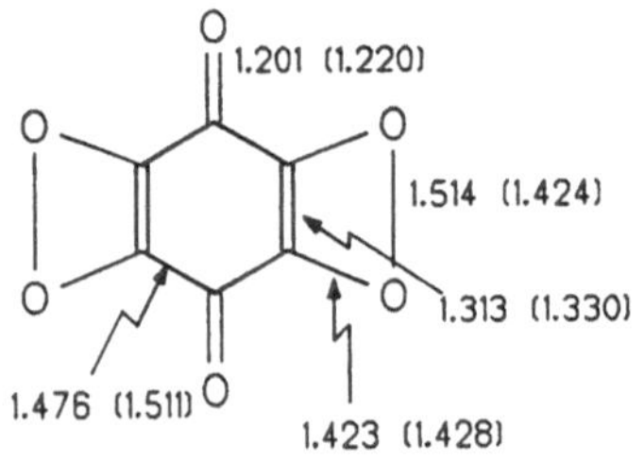

1.201 (1.220)
1.514 (1.424)
1.313 (1.330)
1.476 (1.511)
1.423 (1.428)

148.0 kcal/mol

7. CALCULATION OF ORGANOMETALLIC COMPOUNDS

The examples described in the previous sections all involved molecules which contain only elements of the first and second row of the periodic system of the elements. As mentioned in paragraph 2, the application of _ab initio_ methods to larger systems is limited by the size of the molecule, and the crucial part of the calculations concerns the number of the two-electron integrals $<\mu\nu|\lambda\sigma>$ which constitutes the N^4 "bottle-neck" of all Hartree-Fock based procedures. Thus, it is not only the size of the molecule, i.e. the number of the atoms, but also the **type** of atoms, i.e. the number of electrons which increases with larger atomic number. This is the reason why _ab initio_ calculations have mainly be devoted to the chemistry of first and second row elements. _Ab initio_ calculations of molecules with heavier atoms have in most cased been restricted to diatomics and triatomic systems. However, there is a method available which reduces the computational effort for molecules with heavy atoms to compounds of the same size, but with "light" atoms of the first or second row. In this method only the valence electrons are explicitly treated in a Hartree-Fock calculation analogous to all-electron procedures, while the chemically inert inner core electrons are replaced by a spherical harmonic potential which shields the nuclear attraction. These are called "effective core potentials" (ECP) or "pseudopotentials" [43]. It may be argued that ECP methods can not be considered as genuinely _ab initio_ procedures. We do not think so, because ECP potentials can be derived in the same way as the mathematical functions which serve as basis functions (atomic orbitals) in all-electron _ab initio_ calculations.

ECP methods are very attractive because they mean that _ab initio_ computational chemistry, which has proven a very powerful research tool in the chemistry of "light" elements [8], may extend its scope to _all_ elements of the periodic systems. Since relativistic effects may also be included in the effective core potentials [43,44], even the "very heavy" elements of the sixth and even 7th row of the periodic system can be treated by ECP methods. ECPs have been developed now for nearly all elements [43]. What is needed are tests of the accuracy of the various approaches. In particular, two questions must be answered: (i) What is the best choice of the valence region, i.e. how many electrons should be allowed to be included in the core? (ii) What is the best basis set for the valence electrons after the selection between valence and core electrons has been made? It is the proper balance between (i) and (ii) which decides about the relation between accuracy and computational costs. Unlike all-electron _ab initio_ calculation, there has not been much testing of ECP methods.

We have actively been engaged in systematically testing the various ECP procedures [43]. Here we only show some results [45] of _ab initio_ calculations using ECPs which exhibit the best price/performance ratio [46]. The problem which we addressed concerns the structures and binding energies of donor-acceptor complexes between titanium(IV) compounds as Lewis acid and various organic molecules which serve as Lewis base. Figure 4 shows the geometry-optimized structures of the complexes between ethylene glycole and $TiCl_4$ (**7**) and CH_3TiCl_3 with the methyl group in the axial position (**8a**) and equatorial position (**8b**).

Figure 4. Optimized [45] bond length in Å for **7, 8a, 8b**, using ECP for Ti and 3-21G(d) for other elements.

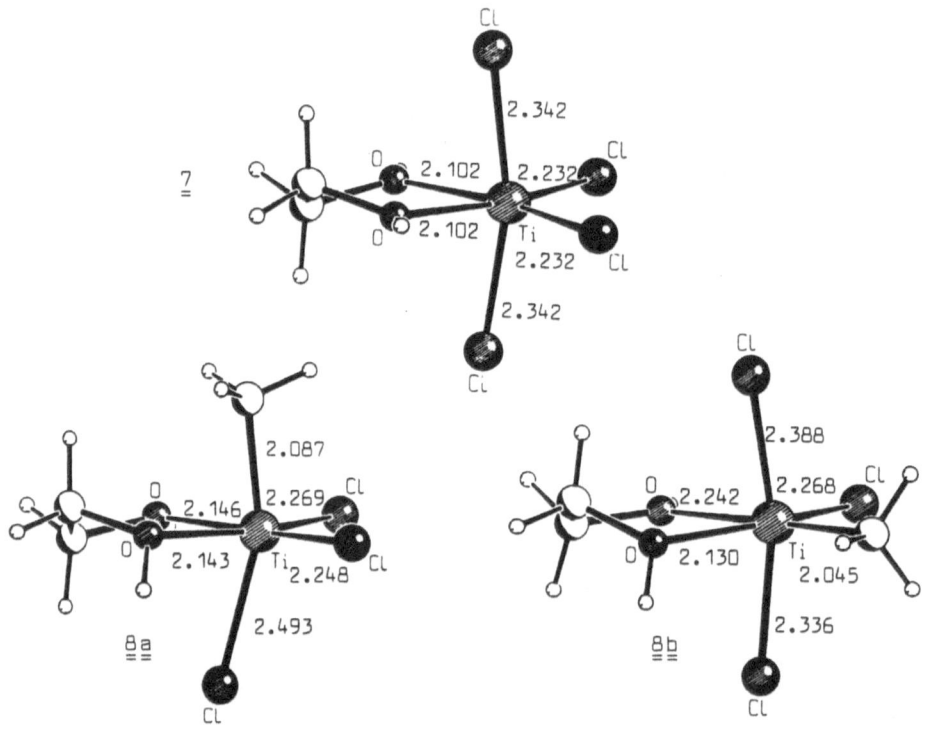

The following binding energies have been computed [45]:

7	---->	$TiCl_4$	+	$(CH_2OH)_2$	+51.0 kcal/mol
8a	---->	CH_3TiCl_3	+	$(CH_2OH)_2$	+44.5 kcal/mol
8b	---->	CH_3TiCl_3	+	$(CH_2OH)_2$	+47.7 kcal/mol

The computational results allow the following conclusions. $TiCl_4$ is a better Lewis acid than CH_3TiCl_3. The methyl group in **8** is preferred in the equatorial position (**8b**) rather than the axial position (**8a**). The axial ligands are bend towards the chelating Lewis base ethylene glycole. The "trans" effect becomes apparent by the calculated Ti-X bond lengths, which are always shorter for Ti-X bonds trans to the glycole group compared to the axial Ti-X bond of the same type (Figure 4). These results are in agreement with experimental data obtained for various transition metal chelate complexes. However, such data are experimentally difficult to obtain, and sometimes it is impossible to study such complexes experimentally. The power of ECP methods may be exemplified by a second example. Figure 5 shows the optimized geometries of the complexes between the unsymmetrical hydroxypropionaldehyde and $TiCl_4$ (**9**) and CH_3TiCl_3, which yields four different isomers **10a, 10b. 10c,** and **10d** [45].

Figure 5. Optimized [45] bond length in Å for complexes **9** , **10a** – **10d** using ECP for Ti and 3-21G(d) for other elements.

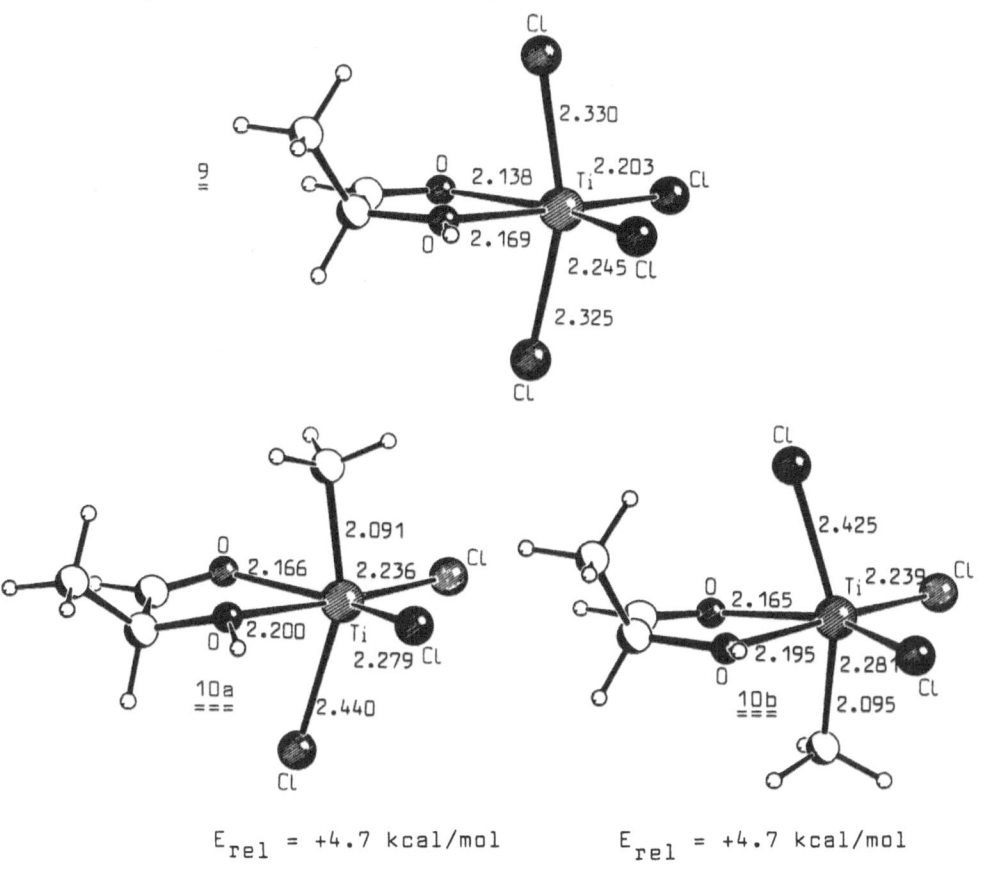

E_{rel} = +4.7 kcal/mol E_{rel} = +4.7 kcal/mol

E_{rel} = 0.0 kcal/mol E_{rel} = +1.7 kcal/mol

The following interaction energies are predicted from the calculations [45]:

9	---->	$TiCl_4$	+	$CH_3C(OH)H-CHO$	+38.0 kcal/mol
10a	---->	CH_3TiCl_4	+	$CH_3C(OH)H-CHO$	+31.7 kcal/mol
10b	---->	CH_3TiCl_4	+	$CH_3C(OH)H-CHO$	+31.7 kcal/mol
10c	---->	CH_3TiCl_4	+	$CH_3C(OH)H-CHO$	+36.4 kcal/mol
10d	---->	CH_3TiCl_4	+	$CH_3C(OH)H-CHO$	+34.7 kcal/mol

Thus, a methyl group is more stable in the trans position to the hydroxy group (**10c**) than trans to the carbonyl group (**10d**). The complexes with axial methyl groups (**10a, 10b**) are least stable.

The theoretical data provide not only insight into the geometrical and electronic structure of the molecules, they may also help to design new effective routes for the synthesis of important compounds. The isomeric forms of **10** are chiral compounds, and the use of alkyl-$TiCl_3$ as chelating agent has been proven very effective for the synthesis of diastereoselective and enantioselective compounds [47]. The knowledge about the most stable isomers of the Lewis acid-base complexes such as **10**, which is difficult to obtain experimentally, may help to design more selective Lewis acids.

8. SUMMARY

In the previous paragraphs, we have described how _ab initio_ methods provide results which may serve to supplement, understand, facilitate, stimulate, and sometimes question experimental chemical studies. The development of quantum chemical methods is an ongoing process which will see much progress in the near future, in particular in the treatment of heavy element molecules. Still, it has already matured to a degree that is has become the most powerful single technique to study chemical phenomena, being predictive and analyzing at the same time. The few examples described above may serve to show how theory and experiment have become two areas of chemical research which often complement one another. The advent of modern supercomputers has helped not only to create faster computations, but to establish a new field in chemistry. Computational Chemistry is now recognized as an important new technique, which constitutes a healthy rivalry with experiment.

ACKNOWLEDGMENTS

The results of our theoretical studies described here have been made possible by the very diligent and careful work of my enthusiastic coworkers which are named as coauthors of the various projects. The continous support of our work by the Deutsche Forschungsgemeinschaft and the Fonds der Chemischen Industrie is gratefully acknowledged. Further support has been provided by the computer companies Convex and Silicon Graphics.

REFERENCES

1 The latest version of the ongoing development in this series is Gaussian 90: M.J. Frisch, M. Head-Gordon, G.W. Trucks, J.B. Foreman, H.B. Schlegel, K. Raghavachari,M.A. Robb, J.S. Binkley, C. Gonzalez, D.J. DeFrees, D.J. Fox, R.A. Whiteside, R. Seeger, C.F. Melius, J. Baker, R. Martin, L.R. Kahn, J.J.P. Stewart, S. Topiol, and J.A. Pople, Gaussian Inc., Pittsburgh, PA 1990.

2 R.D. Amos and J.E. Rice, 'CADPAC: The Cambridge Analytical Derivatives Package', issue 4.0, Cambridge, 1987.

3 The original version of GAMESS was developed by the staff of the NRCC: M. Dupuis, D. Spangler, and J.J. Wendolowski, National Resources for Computations in Chemistry, 1980. Since then, several versions have been developed in different groups such as the "Daresbury version" by M. Guest and the "Dakota version" by M. Schmidt and S. Elbert.

4 M.J.S. Dewar and W. Thiel, J. Am. Chem. Soc. **99**, 4899, 4907 (1977).

5 M.J.S. Dewar, E.G. Zoebisch, E.F. Healy, and J.J.P. Stewart, J. Am. Chem. Soc. **107**, 3902 (1985).

6 J.J.P. Stewart, J. Comput. Chem. **10**, 209, 221 (1989).

7 An excellent overview over the state-of-the-art in computational chemistry is given in: "Reviews in Computational Chemistry", K.B. Lipkowitz and D.B. Boyd (Eds), VCH Publishers, New York, 1990.

8 W.J. Hehre, L. Radom, P.v.R. Schleyer and J.A. Pople, "Ab Initio Molecular Orbital Theory", Wiley, New York, 1986.

9 K.O. Christe and W. Sawodny, Inorg. Chem. **8**, 212 (1969).

10 G. Frenking and W. Koch, Inorg. Chem. **29**, 4513 (1990).

11 J. Gillespie and M.J. Morton, Inorg. Chem. **9**, 811 (1970).

.12 (a) B.D. Joshi and K. Morokuma, J. Am. Chem. Soc. **101**, 1714 (1979). (b) S. Bhattacharjee, A.B. Sannigrahi and D.C. Mukherjee, Indian J. Chem. **22A**, 1001 (1983). (c) S. Bhattacharjee and A.B. Sannigrahi, Indian J. Chem. **23A**, 285 (1984). (d) R.L. DeKock, C.P. Jasperse, D.T. Dao, J.H. Bieda, J. F. Liebman, J. Fluorine Chem. **22**, 575 (1984). (e) J.K. Burdett and C.J. Marsden, New J. Chem. **12**, 797 (1988).

13 Apart from the errors introduced by the approximations in the computational procedure (correlation energy, basis set deficiency), the calculation of the vibrational spectra using second derivatives is subject to the harmonic approximation which in most cases yields frequencies which are 5 - 15% too high, s. ref. 8.

14 J.A. Pople, M.J. Frisch, B.T. Luke and J.S. Binkley, Int. J. Quant. Chem. Quantum Chem. Symp. **17**, 307 (1983).

15 W.J. Hehre and J.A. Pople, J. Am. Chem. Soc.**92**, 2191 (1970).

16 Huber, K.P.; Herzberg, G. "Constants of Diatomic Molecules"; Van Nostrand Reinhold, New York, 1979.

17 K. Dehnicke, H. Aeissen, M. Kölmel, and J. Strähle, Angew. Chem. **89**, 569 (1977); Angew. Chem. Int. Ed. Engl. **16**, 545 (1977).

18 M. Brumm, W. Koch, and G. Frenking, Chem. Phys. Lett., in press.

19 E.A. Meyers and W.N. Lipscomb, Acta Crystallogr. **8**, 853 (1955).

20 N. Bartlett, J. Passmore and E.J. Wells, Chem. Commun. 213 (1966).

21 R. Schulz, A. Schweig, K. Hartke, J. Köster, J. Am. Chem. Soc. 105 (1983) 4519.

22 M. Rodler and A. Bauder, Chem. Phys. Lett. **114**, 575 (1985).

23 (a) F. Diehl, H. Meyer, A. Schweig, B.A. Hess, Jr. and J. Fabian, J. Am. Chem. Soc. **111**, 7651 (1989). (b) R. Janoschek, B.A. Hess, Jr. and J. Fabian, Z. Chem. **30**, 176 (1990).

24 As a curiosity, the correct stability order **3b** > **3a** is predicted at the simple STO-3G, STO-3G+6D and 4-31G+6D level using STO-3G optimized geometries: R.C. Haddon, S.R. Wasserman, F. Wudl, and G.R.J. Williams, J. Am. Chem. Soc. 102 (1980) 6687.

25 J.D. Goddard, J. Comput. Chem. **8**, 389 (1987).

26 See table 6.70 in ref. 8.

27 V. Jonas and G. Frenking, Chem. Phys. Lett. **177**, 175 (1991).

28 A. Reed and P.v.R. Schleyer, J. Am. Chem. Soc. **112**, 1434 (1990).

29 Greenwood, N.N.; Earnshaw, A. "Chemistry of the Elements"; Pergamon Ress, Oxford, 1984, p. 985.

30 The structural data of the unit cell were kindly given to us by J. Strähle and F. Weller.

31 (a) A. Szabo and N.S. Ostlund, "Modern Quantum Chemistry", Macmillan, New York, 1989. (b) R. Daudel, G. Leroy, D. Peeters, and M. Sana, "Quantum Chemistry", Wiley, New York, 1983. (c) G. Naray-Szabo, P. R. Surjan, and J.G. Angyan, "Applied Quantum Chemistry", Reidel, Dordrecht, 1987.

32 (a) D.R. Hartree, Proc. Cambr. Phil. Soc. **24**, 89 (1927). (b) V. Fock, Z. Phys. **61**, 126 (1930).

33 M. Born and R. J. Oppenheimer, Ann. Phys. **84**, 457 (1927).

34 (a) C.C.J. Roothaan, Rev. Mod. Phys. **23**, 69 (1951). (b) G.G. Hall, Proc. Roy. Soc. **A205**, 541 (1951).

35 For a discussion of the approximations in modern quantum chemistry see: H.F. Schaefer (Ed) "Modern Theoretical Chemistry", (a) Vol. 3 and (b) Vol. 4, Plenum, New York, 1977.

36 (a) C. Møller and M.S. Plesset, Phys. rev. **46**, 618 (1934). (b) J.S. Binkley and J.A. Pople, Int. J. Quantum Chem. **9**, 229 (1975).

37 I. Shavitt, in ref. 35a, p. 189.

38 G. Maier, H.-P. Reisenauer, J. Schrot, R. Janoschek, Angew. Chem. **102**, 1475 (1990); Angew. Chem. Int. Ed. Engl. **29**, 1464 (1990).

39 D. Sülzle, N. Beye, E. Fanghänel, H. Schwarz, Chem. Ber. **122**, 2411 (1989).

40 G. Frenking, Angew. Chem. **102**, 1516 (1990); Angew. Chem. Int. Ed. Engl. **29**, 1410 (1990).

41 H.-P. Reisenauer, G. Maier, personal communication.

42 H.-P. Reisenauer, G. Maier, V. Jonas, G. Frenking, to be published.

43 (a) P.J. Hay and W.R. Wadt, J. Chem. Phys. **82**, 270, 284, 299 (1985). (b) Y. Sakai, E. Miyoshi, M. Klobukobski, and S. Huzinaga, J. Comput. Chem. **8**, 256 (1987). (c) U. Wedig, M. Dolg, H. Stoll, and H. Preuss, in "Quantum Chemistry: The Challenge of Transition Metals and Coordination Chemistry", A. Veillard (Ed), Reidel, Dordrecht, 1986.

44 P. Pyykkö, Chem. Rev. **88**, 563 (1988), and further references cited therein.

45 V. Jonas, M.T. Reetz, and G. Frenking, manuscript in preparation.

46 V. Jonas and G. Frenking, manuscript in preparation.

47 M.T. Reetz, "Organotitanium Reagents in Organic Synthesis", Spinger, Heidelberg, 1986.

48 K.O. Christe and W. Maya, Inorg. Chem. **8**, 1253 (1969).

Parallel Processing and Computational Chemistry

Paul Weiner, PhD

Amber Systems, Inc., 28 Tower Street, Somerville, MA 02143, USA

With the increasing use of such methods as dynamic simulations, free energy perturbation and large molecule ab-initio calculations, as well as the use of very elaborate graphical display programs, chemical researchers find increasing frustration in the use of conventional computers to solve their problems. These machines are either too slow or too expensive to be practical for everyday use. However, the advent of parallel computer architectures and parallel algorithms offers a solution to this problem.

Before discussing the types of parallel computers and the programming requirements to run codes efficiently on them, it is important to define the basic idea behind parallel computation. In Figure 1, an example of a loop that can be run in parallel is given. The fact that all tasks are independent and can be executed in any order implies that there can be no data dependencies between any of the tasks. Any data dependency will result in the need for a particular task to be run before another one and the tasks can no longer be sent independently to different processors. Thus a major programming problem of converting code from a single to a multiple processor computer is the removal of data dependencies from the computationally intensive loops. There are many tools, which will be discussed later, to assist in this process.

There are two common types of parallel computers – shared memory and distributed memory machines. The shared memory computers have a central memory that is shared by all processors. This is the easiest type to program since the data is available to all processors. However, the programmer must worry about the data integrity. If there is a loop of code that has different iterations executed across different processors and there is an array that uses values calculated in a previous iteration, the code may give incorrect answers. This is due to data dependencies in the loop. The programmer must find a way to avoid this problem.

The distributed memory computers, such as the hypercube, have several nodes loosely coupled. Each node has a processor, memory, and some communication hardware and operates independently. The nodes work

U. Harms (Ed.)
Supercomputer and Chemistry 2
© Springer-Verlag Berlin Heidelberg 1991

134

```
do I = 1, n
        TASKᵢ
enddo
```

Where Task₁ , . . . , Task n
Are *Independent.* **They may be executed in any order:**

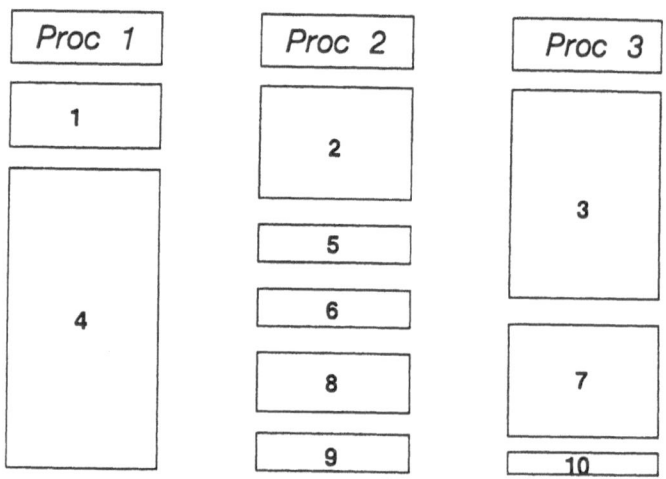

Figure 1 – The Basic Unit of Parallel Computation

together by sending and receiving messages. These systems can scale up relatively cheaply to hundreds or thousands of processors. These systems can achieve very impressive speeds. However, if there is a need for the nodes to frequently communicate, the performance drops quickly. In addition, these machines can be difficult to program when there is a need for frequent communication between nodes. The program must be modified to include code both to send and to receive messages. After this has been done, the code will often need to be modified to be run on a serial or a shared memory parallel computer.

There are many opportunities for both high-level or coarse-grain parallelism and loop-level or fine-grain parallelism in chemistry codes. In molecular mechanics codes, these opportunities occur in the computation of the nonbonded pairwise interactions. In ab-initio codes, the integral and derivative calculations, the Fock matrix construction, as well as some parts of the eigenvalue problem may be computed concurrently. Also, most of the post-SCF calculations involve inherently parallel matrix operations. The algorithms to render atoms, bonds, and orbitals are highly parallelizable.

To exploit the parallelism that exits in these codes, special parallel code must be written by the user or a compiler must be used that recognizes opportunities for parallelism in the serial code. In a shared memory computer, fine-grained parallelism can be handled by the compiler automatically. Simple data dependencies in a loop will be detected by the compiler and do not prevent parallelism. Figure 2 gives an example of a loop that has a simple data dependency, but which can still be run concurrently simply by waiting for the required values of F(i) to be computed before going on with the calculation. All codes can take advantage of their fine–grain parallelism merely by being compiled.

Medium–grain parallelism involves several loops or many operations within a single loop. Since many operations are carried out, it is often too difficult for the compiler to analyze the code and to locate data dependencies. However, the programmer understands the algorithm and often knows whether or not data dependencies occur. In this case, it is possible to give directives to the

Tools: Compiler (Automatic)

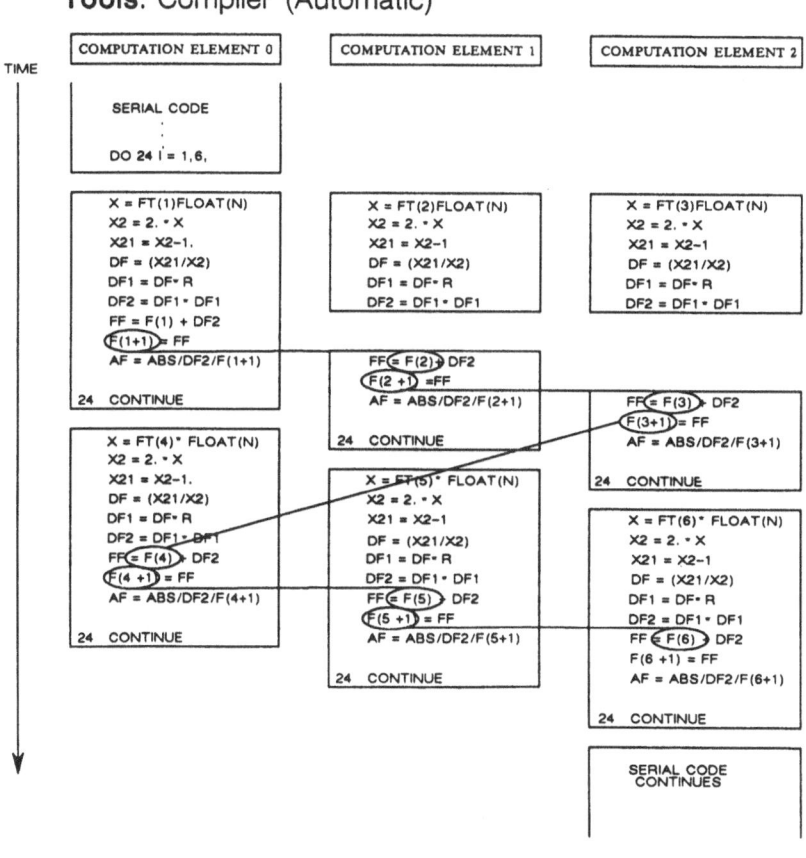

Figure 2 – Fine-Grain Parallelism

compiler that tell it to go ahead and make the loop run concurrently. In Figure 3, the programmer knows that the three calls to molecular mechanics routines are all independent and can run on separate processors. The CVD$CNCALL directive tells the compiler to run this loop concurrently. The second example uses a directive to tell the compiler to ignore dependency checking because the index j(i) has different values for all values of i in the do loop.

Coarse-grain parallelism involves very high level parallelism and often occurs in many nested loops that contain complex subroutine calls. This type of parallelism usually involves human intervention; however, it is the most effective form of parallelism. There are two powerful tools developed by Alliant to help the user take full advantage of this form of parallelism. Both tools must address the problem of data dependencies in the code.

Tools: Directives

```
CVD$ CNCALL
              do inter = 1, 3
                  if (inter.eq. 1) call bond
                  if (inter.eq. 2) call theta
                  if (inter.eq. 3) call phi
              end do

CVD$ nodepchk
              do 13 i = 1, u
          13      a(j(i)) = a (j(i)) + s
```

Figure 3 – Medium-grain Parallelism

The first tool is called FAST and is illustrated in Figure 4. It performs an inter-procedural analysis to detect dependencies in the code and then automatically rewrites the code to remove the dependencies. Each processor is given a unique copy of any array that has a data dependency and the results of each processor are combined after the last iteration is completed.

This method has the disadvantage of requiring a complex interprocedural analysis and the algorithm is difficult to generalize for arbitrary Fortran constructs. There can also be load balancing problems when the loops do not require approximately the same amount of work. In this case, it is possible for several processors on a complex to remain idle while the remaining processors finish up their computations. In an effort to correct these deficiencies, a new algorithm, FASTf, was developed at Alliant. This algorithm is based on forking of processes and does not require restructuring of code to protect the global data. The shared data and communications between the processes occurs via the shared memory regions. The user only needs to add directives to the code (for the compiler) in order to run a section of code concurrently. Figure 5 shows an example of the directives required for a typical molecular mechanics kernal.

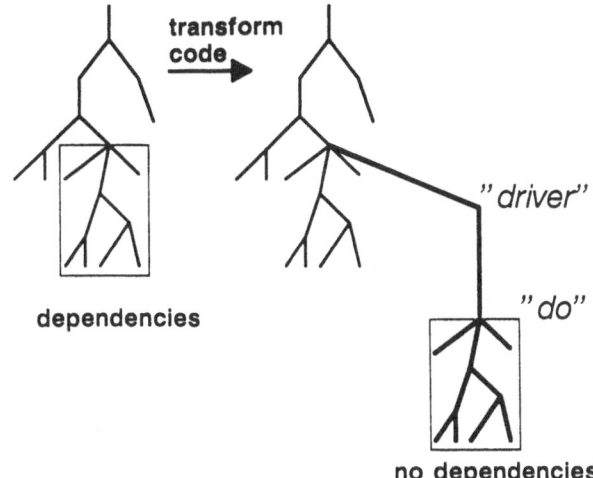

transform
code →

"driver"

"do"

dependencies

no dependencies

Figure 4 – Tools For Parallelization: Fast
Method

- – Perform inter-procedural analysis to detect dependencies

- – Rewrite code to remove dependencies

Example for a Typical Molecular Mechanics Kernel:

```
CVD$  DATA_DEFS
      preliminary work; initialization, etc.

CVD$  PARALLEL REGION
c  non-bonded terms
   do i = 1,nat
      a little work done here
      do j = 1,i
CVD$  PDO
              here comes the bulk of the work
              to compute gradients and energy
CVD$  END PDO
      enddo
   enddo
c  coulombic-terms
   some preparation
   do i = 1,nat
      a little work done here
      do j = 1,i
CVD$  PDO
              here comes the bulk of the work
              to compute gradients and energy
CVD$  END PDO
      enddo
   enddo
c  bonded terms
   some preparation
   do i = 1,list_bonded
      a little work done here
```

```
CVD$  PDO
              here comes the bulk of the work
              to compute gradients and energy
CVD$  END PDO
      enddo
c  bending-terms
   some preparation
   do i = 1,list_bending
CVD$  PDO
              here comes the bulk of the work
              to compute gradients and energy
CVD$  END PDO
      enddo
c  torsional-terms
   some preparation
   do i = 1,list_torsional
CVD$  PDO
              here comes the bulk of the work
              to compute gradients and energy
CVD$  END PDO
      enddo
CVD$  END PARALLEL REGION
CVD$  JOIN gradients $DIM 3,natom
CVD$  JOIN etotal
```

Figure 5 – Tools For Parallelization: Fastf

138

The directives are used to mark the beginning and end of the code segments that are to be executed in parallel by the different processes and are replaced by FASTf with calls to subroutines that handle forking and joining of processes. (These directives are explained in detail in the Alliant *Focus on Chemistry* brochure by the developer of the technique – Roberto Gomperts, PhD.)

The advantages of FASTf are the ability of a job to use multiple processors only when it needs to and the fact that processors are available for other tasks when they are not needed. The disadvantage of FASTf is its inability to handle very fine-grain parallelism well. In this case, forking and joining overhead can offset gains from running a parallel execution of the code.

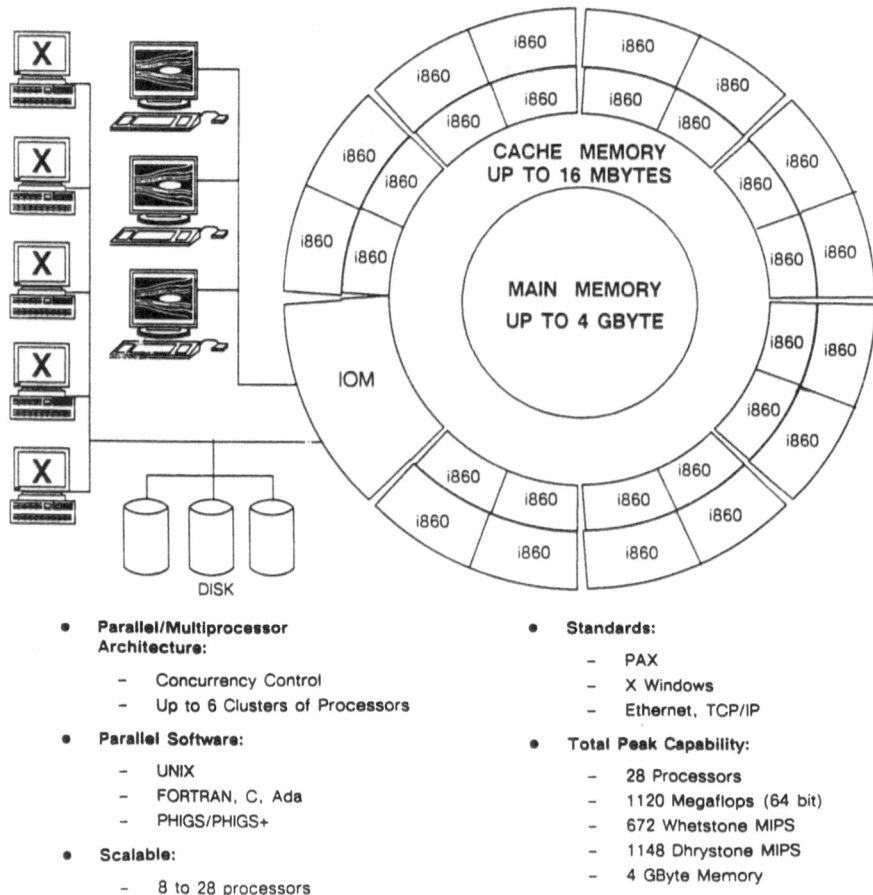

- **Parallel/Multiprocessor Architecture:**
  - Concurrency Control
  - Up to 6 Clusters of Processors

- **Parallel Software:**
  - UNIX
  - FORTRAN, C, Ada
  - PHIGS/PHIGS+

- **Scalable:**
  - 8 to 28 processors
  - 4 processors per module

- **Standards:**
  - PAX
  - X Windows
  - Ethernet, TCP/IP

- **Total Peak Capability:**
  - 28 Processors
  - 1120 Megaflops (64 bit)
  - 672 Whetstone MIPS
  - 1148 Dhrystone MIPS
  - 4 GByte Memory

Figure 6 – FX/2800 Architecture – Shared Memory Parallel Computer

A combination of all of these methods are used on computational chemistry codes to decrease the time–to–solution of a single job, as well as to increase the throughput or to decrease the time–to–solution of multiple jobs.

Examples of several types of parallelism will now be given on the Alliant FX/2800. This computer is a shared memory parallel computer with a maximum of 28 I860 processors. The processors can be combined to form a complex, for optimum time–to–solution for a particular job, or run individually in detached mode, for high throughput. Combinations of these are also allowed. Figure 6 lists some of the capabilities of the Alliant.

Figure 7 shows the speedup achieved on a molecular mechanics code, AMBER 3+ (developed by Chandra Singh and Amber Systems, Inc. and derived from AMBER 3.0 by Chandra Singh, Paul Weiner, and Peter Kollman), by taking advantage of medium–grain parallelism. Directives were used to make the non–bonded routine run concurrently. The rest of the code exploited the fine-grain parallelism that the compiler detected. The code achieved a speedup of 4.2X in going from 1 to 6 processors.

Figure 8 shows the speedup achieved on an ab–initio code. These codes are much more difficult to parallelize than the molecular mechanics codes and typically require the use of coarse–grain parallelism tools. This code has been parallelized with FASTf, using forking, and achieved a 4.1X speedup in going from 1 to 8 processors.

# of Processors	Time (sec.)
1	707.5
2	455.0
4	237.8
6	167.4

1121 Atoms, 10Å NB cutoff, 100 steps of minimization.

Figure 7– Single Job Comparison of Parallel and Non-parallel Amber 3+ – FX/2800

140

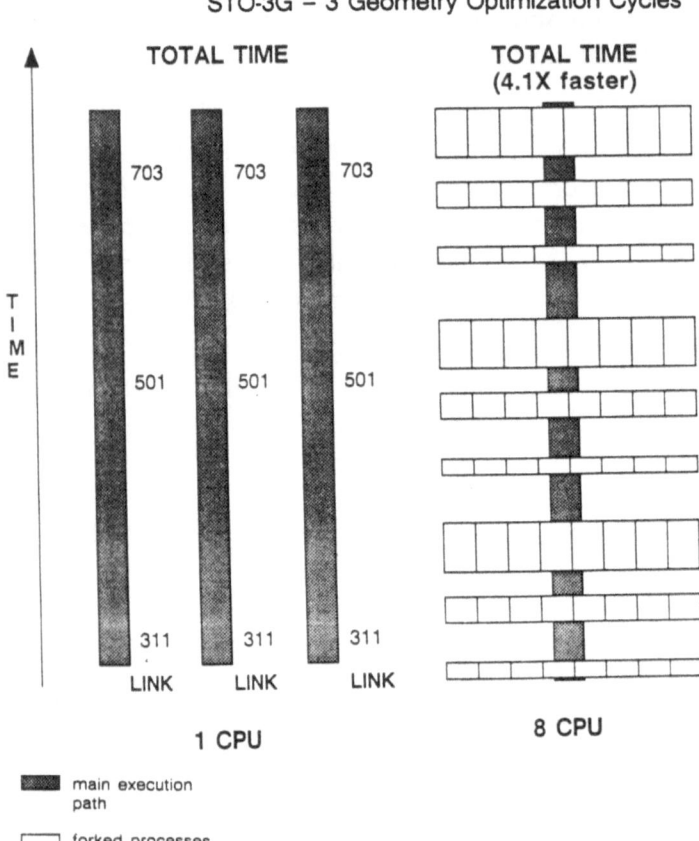

23 atoms, 68 basis functions
STO-3G – 3 Geometry Optimization Cycles

Figure 8 – Single Job Comparison of Parallel and Non-parallel
Ab-Initio – FX/2800

These examples, both show how parallelism can help achieve a faster time–to–solution for a single job. Another use of parallelism is to help achieve a faster throughput or time–to–solution of multiple jobs. Figures 9, 10, and 11 give a mix of computational chemistry jobs. Table 1 shows the results of running each job individually ("Time To Solution" column) and the time for running each job when all jobs are run at once ("Under Load" column). If all jobs were run one after another, they would take 4 hours and 50 minutes. By running all jobs simultaneously, they only take a total of 1 hour to run.

Molecular Mechanics – Discover 2.60

Test1	–	1876 atoms 150 steps of Minimization 225 steps of Dynamics
Test2	–	2073 atoms 800 steps of Minimization
Test3	–	2073 atoms 300 steps of Minimization
Test4	–	3082 atoms 600 steps of Dynamics

Figure 9 – Computational Chemistry Load Test – Simultaneous Jobs

Table 1 – Computational Chemistry Load Test – Simultaneous Jobs

Application/Test	# Processors	Time To Solution	Under Load
Discover			
Test1	4	30:58	44:23
Test2	4	34:28	50:56
Test3	4	31:14	44:51
Test4	4	31:23	46:36
Mopac			
Test1	1	34:57	59:24
Test2	1	37:33	58:04
Test3	1	26:17	46:07
Ab-initio			
Test1	8	39:00	1:00:35
Test2	4	24:22	39:23
Total	31	4:50:12	1:00:38

Times are wall-clock times in hour:minutes:seconds

FX/2828 – September, 1990

Semi-Empirical – MOPAC 5.0

Test1   –   96 atoms (C37 H48 N4 07)
                 1 Cycle of Nllsq using AM1

Test2   –   20 atoms (C6 H9 N 03 S)
                 300 Cycles of Nllsq using AM1

Test3   –   24 atoms (C14 H10)
                 90 Cycles of Nllsq using MNDO

**Figure 10 – Computational Chemistry Load Test –
Simultaneous Jobs**

Ab-Initio

Test1   –   18 atoms (C5 H10 N O CL)
                 $6\text{-}31G^*$, 144 basis functions
                 600 MB disk
                 Single point RHF

Test2   –   10 atoms (C2 H6 O S)
                 $6\text{-}31G^*$, 76 basis functions
                 7 MB disk
                 Direct SCF (RHF) Geometry Opt.

**Figure 11 – Computational Chemistry Load Test –
Simultaneous Jobs**

Both forms of parallelism can also be combined to run multiple jobs, all working on different aspects of the same research problem, at the same time. An example of this, illustrated in Figure 12 on an Alliant FX/80, is a set of tasks consisting of AMBER 3+, AMBER 3+ analysis, and the GRAMPS graphics display program. These programs are traditionally run in serial mode – batch processing of AMBER 3+, followed by analysis and graphics post–processing. Paul Weiner, Steve Gallion, Rosario Caltabiano, and T.J. O'Donnell wrote a program that uses UNIX shared memory calls for synchronization and data passing between the applications. The use of intermediate disk files, which slows execution, is avoided. AMBER 3+ passes the results of the dynamics simulation to the analysis program. After analyzing the results, this program passes the results to GRAMPS, which then renders the molecule in colors that represent values computed in the analysis program. AMBER 3+ can be computing the next dynamics point while the analysis and graphics programs are working.

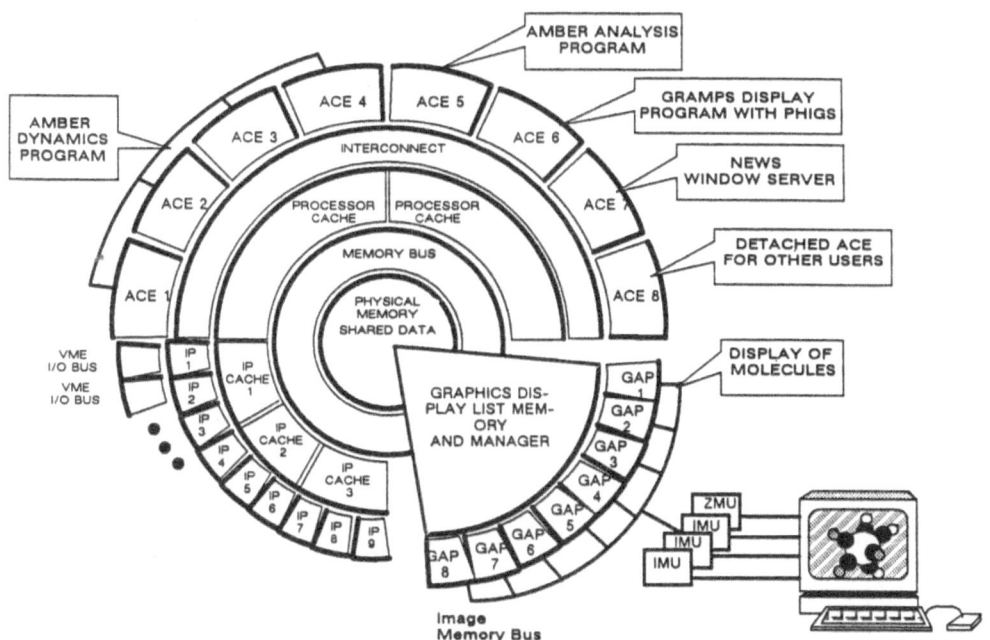

**Figure 12 – Throughput: Multiple Chemistry Codes
Solving the Same Problem**

For this process to work, it is important for all steps to take approximately the same time. This is normally not true and requires that the computationally intensive steps be run in parallel. The computer must also be able to efficiently run several different jobs at the same time to maintain interactive performance. By achieving these goals, the user can interactively explore conformational space for a small molecule by using a combination of minimization and dynamics steps.

In the future, there is a possibility of combining multiple forms of parallelism within a single program. Both FAST and FASTf could be combined and have different subroutines run concurrently across different numbers of processors. This will help achieve both optimal time-to- solution for a single job and high throughput for multiple jobs. By running multiple programs at once, it may be possible to interactively explore conformational space and reaction pathways for systems of chemical interest.

The Direct IGLO Method for the Calculation of NMR Chemical Shifts with the Program TURBOMOLE

Ulrich Meier

CONVEX Computer GmbH, Hansaallee 201, W-4000 Düsseldorf 11, FRG

Abstract: The IGLO method for the "ab initio" calculation of NMR chemical shifts is programmed using the two electron integral features of the program package TURBOMOLE. A data flow optimized implementation on supercomputers makes calculations for large ($N \leqslant 1000$ Gaussian basis functions) possible by replacing disk I/O by direct and parallel calculations.

In this lecture we report on further developments [1] of the Bochum IGLO (Individual Gauge for Localized Orbitals) program [2,3,4] using the two electron integral handling of the program package TURBOMOLE [5,6]. When implemented on supercomputers it permits the "ab initio" calculation of NMR chemical shifts for large molecules, where "large" means up to $N \approx 1000$ basis functions $K \approx 100$ nuclei. This will be achieved by:

i) "HONDO-type" calculation of the one-electron integrals

ii) two-electron integral processing as in the program TURBOMOLE

iii) direct and parallel computation of the individually gauged one-electron-operators and the $K_o^+(j)$ operators.

To stress the computational problems of solving the IGLO equation

U. Harms (Ed.)
Supercomputer and Chemistry 2
© Springer-Verlag Berlin Heidelberg 1991

we briefly repeat the working equations of the IGLO formalism. For more details we refer the reader to the literature [2,3,4].

In the IGLO formalism the coupled Hartree-Fock (CHF) perturbation equations are written in terms of localized orbitals (LMO's) according to the criterion of Boys [7].

The perturbed molecular orbitals Ψ_k the projectors $\tilde{P}_k$ and the Fockoperators $\tilde{F}_k$ are expanded in powers of the perturbation parameter
- the magnetic field strength λ .

$$\Psi_k = \Psi_{ko} + i\lambda\Psi_{k1} + \lambda^2\Psi_{k2} + \ldots \qquad (1)$$

$$P_k = P_o + i\lambda P_{k1} + \lambda^2 P_{k2} + \ldots \qquad (2)$$

$$F_k = F_o + i\lambda F_{k1} + \lambda^2 F_{k2} + \ldots \qquad (3)$$

$$\tilde{P}_{k1} = \sum_{l=1}^{M}\left\{|\Psi_{l1}><\Psi_{lo}| - |\Psi_{lo}><\Psi_{l1}|\right.$$
$$\left. + [\Psi_{lo}><\Psi_{lo}, \lambda_k - \lambda_l] \right\} \qquad (4)$$

$$\tilde{F}_{k1} = \tilde{h}_{k1} - \sum_{j}^{M}\tilde{K}_l(j) + \sum_{j}^{M}[\tilde{K}_o^+(j), \lambda_j - \lambda_k] \qquad (5)$$

with $\lambda_k(r)$ being the gauge function.

$$\lambda_k(\underline{r}) = \lambda\frac{e}{2ch}(\underline{R}_k \times \underline{B})\,\underline{r} \qquad (6)$$

M: number of doubly occupied orbitals.

The first order perturbation correction of the MO's for each direction of the magnetic field is determined by

$$\left\{(1-P_o)\tilde{F}_{k1} - \tilde{P}_{k1} F_o\right\}\Psi_{ko} + (1-P_o) F_o\Psi_{k1} = 0 \qquad (7)$$

The exchange operators are defined via their matrix elements. They are computed in the AO basis from a density matrix D and the appropiate combination of two-electron integrals.

$\tilde{K}_0^+$ (j) - operator:

$$< \mu | \tilde{K}_0^+ (j) | \nu > = \sum_{\rho \geq \sigma} D_{\rho\sigma}^{jj} \left\{ (\mu\rho|\nu\sigma) + (\mu\sigma|\nu\rho) \right\} \qquad (8)$$

with

$$D_{\rho\sigma}^{jj} = c_{j\rho}^{\sigma} c_{j\sigma}^{\sigma} (1 - \frac{1}{2}\delta_{\rho\sigma}) \qquad (9)$$

and the localized molecular orbitals (LMO)

$$\varphi_k^{\sigma} = \sum_{\mu=1}^{N} c_{k\mu}^{\sigma} x_{\mu} \qquad (10)$$

The $\tilde{K}_1^-$ -operator is needed during the iterative solution of the IGLO equations. It is defined as

$$< \mu | \tilde{K}_1^- | \nu > = \sum_{j=1}^{M} < \mu | \tilde{K}_1^- (j) | \nu > \qquad (11)$$

$$< \mu | \tilde{K}_1^- | \nu > = \sum_{\rho > \sigma} D_{\rho\sigma}^{1} \left\{ (\mu\rho|\nu\sigma) - (\mu\sigma|\nu\rho) \right\} \qquad (12)$$

$$D_{\rho\sigma} = \sum_{j=1} \frac{1}{2} (c_{j\rho}^{\sigma} c_{j\sigma}^{1} - c_{j\rho}^{1} c_{j\sigma}^{\sigma}) (1 - \frac{1}{2}\delta_{\rho\sigma}) \qquad (13)$$

Since the perturbed orbitals enter equation (7) through the Fock-operators, this equation is solved iterativly until selfconsis-tency is reached.

i) one-electron integrals

The most difficult one-electron integral forming the $\tilde{h}_{kl}$ of equation (5) is of the form

$$< \mu | \frac{(x - x_c)}{|r - c|^3} | \nu > \qquad (14)$$

It can be reduced to

$$< \mu | \frac{(x - x_c)}{|r - c|^3} | \nu > = < \mu | (\frac{\partial}{\partial x} \frac{1}{|r - c|}) | \nu >$$

$$= - < \frac{\partial\mu}{\partial x} | \frac{1}{|r - c|} | \nu > + < \mu | \frac{1}{|r - c|} \frac{\partial\nu}{\partial x} > \qquad (15)$$

Differentiating a Cartesian Gaussian type function (CGTF) X_u yields a linear combination of CGTF's. From the HONDO framework [8] it is known that

$$\langle \rho | \frac{1}{|r-c|} | \sigma \rangle = \sum_{k=o}^{L} W_k \, I_x \, (t_k) \, I_y \, (t_k) \, I_z \, (t_k) \qquad (16)$$

where L depends on the degrees of polynomials in ρ and σ , t_k and W_k are roots and weights corresponding to a Rys polynomial of degree L.

I_x , I_y and I_z are overlap type integrals depending only on X_c , Y_c or Z_c . All types of one electron integrals for the IGLO equations can be expressed by similar algebraic manipulations as linear combinations of simple to calculate overlap integrals.

ii) two-electron integral handling

From equation (5) it is obvious that the calculation of the $\widetilde{K_o^+}(j)$ operators is the time determining step for larger molecules, requiring $\sim M*N**4$ operations. The loop over M doubly occupied orbitals is efficiently vectorized.

The features of the program TURBOMOLE can be used for the calculation of the exchange operators as well.

ii.1)
Only those integrals that are frequently used or difficult to calculate are stored on disk. All other integrals are recalculated (whenever needed).The usage of peripheral resources is controlled by user defined keys.

ii.2)

It's possible to apply an effective estimation of the contribution
of an integral to the 1st order Fock operator.

The maximum contribution of all integrals $(\mu\nu|\rho\sigma)$ [1] within a
batch $(MN|RS)$ is:

$$\text{CON } (MN,RS) = \max \left\{ D_{MR}, D_{MS}, D_{NR}, D_{NS} \right\} * \text{EST } (MN,RS) \qquad (17)$$

with different expressions for the density matrixes:

For the $K_o^+(j)$ - operators

$$D_{MR} = \max \left\{ |D_{\mu\rho}^{jj}| \right\}, \; j = j_{\text{first}} \cdots j_{\text{last}} \; , \; \mu EM, \; \rho ER \; ^{(2)} \qquad (18)$$

and for the K_1^- operators

$$D_{MR} = \max \left\{ |\Delta D_{\mu\rho}^{1x}|, |\Delta D_{\mu\sigma}^{1y}|, |\Delta D_{\mu\sigma}^{1z}| \right\}, \; \mu EM, \; \rho ER \qquad (19)$$

$$\text{EST}(MN,RS) = \text{EST}(MN) \; \text{EST}(RS) \qquad (20)$$

$$\text{EST}(MN) = \max \left\{ Q_{\mu\nu} \right\}, \mu EM, \; \nu EN \qquad (21)$$

$$Q_{uv} = \sqrt{(\mu\nu|\mu\nu)} \qquad (22)$$

The two electron integrals satisfy a Schwarz' inequality [5,9,10]

$$(\mu\nu|\rho\sigma) \leq Q_{\mu\nu} \; Q_{\rho\sigma} \qquad (23)$$

If CON(MN,RS) is less than a given threshold the computation of
the hole batch (MN,RS) of integrals is skipped.

ii.3)

For the iterative determination of the three components of the
K_1^- operators it is possible to apply the minimizations of the
differential density by linear combination as proposed by Ahlrichs

$$\Delta \underline{\underline{D}}_1^{-(n)} = \underline{\underline{D}}_1^{-(n)} - \sum_{j=1}^{n-1} x_j^{(n)} \Delta \underline{\underline{D}}_1^{-(j)} \tag{24}$$

n, j are iteration counts

$$\Delta \underline{\underline{K}}_1^{-(n)} = \underline{\underline{K}}_1^{-(n)} (\Delta \underline{\underline{D}}_1^{-(u)}) \tag{25}$$

$$\underline{\underline{K}}_1^{-(n)} = \Delta \underline{\underline{K}}_1^{-(n)} + \sum_{j=1}^{n-1} x_j^{(n)} \Delta \underline{\underline{K}}_1^{-(j)} \tag{26}$$

The condition

$$tr \left\{ \Delta \underline{\underline{D}}_1^{-(n)} \ \Delta \underline{\underline{D}}_1^{-(n)} \right\} = min \tag{27}$$

leads to equation

$$\underline{\underline{A}} \ \underline{x} = \underline{b} \tag{28}$$

with

$$A_{ij} = tr \left\{ \Delta \underline{\underline{D}}_1^{-(i)} \ \Delta \underline{\underline{D}}_1^{-(j)} \right\} \tag{29}$$

$$b_i = tr \left\{ \Delta \underline{\underline{D}}_1^{-(i)} \ \Delta \underline{\underline{D}}_1^{-(n)} \right\} \tag{30}$$

as in [5] equation (28) is solved with error control.

The effect of integral estimation and minimization of the differential density on the computation time is shown in table [1] for a calculation of an N-Methylacetanilid (N=308, M=40, K=21) molecule.

---

(1) $\mu\nu\rho\sigma$ AO - indices

    MNRS   shells with μEM, νEN etc.

    ijkl   LMO's (occupied)

    abcd   virtuell (unoccupied) MO's

(2) j_{first}, j_{last} is first/last orbital index of the exchange operators which can be kept in the main memory for one sweep through the integrals.

Table 1 - Impact of effective integral estimation and linear combination of the differential density on the CPU time per iteration for a calculation of N-Methylacetanilid (basis ii[a], N=308, M=40, K=21)

	DIGLO		DSCF	
iteration	d-dens Norm[b]	CPU-time/s[c]	d-dens Norm	CPU-time/s[c]
1	0.30E-1	8622	0.46E+1	8934
2	0.33E-2	7413	0.16E+1	7915
3	0.67E-3	6208	0.50E+0	7417
4	0.13E-3	4877	0.19E+0	6942
5	0.23E-4	3449	0.40E-1	6228
6	0.40E-5	2245	0.11E-1	5522
7	0.25E-5	1798	0.23E-2	4831
8	0.30E-5	1955	0.61E-2	4147
9	0.29E-5	1899[d]	0.15E-3	3486
10	0.19E-5	1744	0.44E-4	2928
11	0.10E-13	304	0.10E-4	2256
12			0.80E-5	2054
13			0.51E-5	1915
14			0.44E-5	1834
15			0.41E-5	1823[d]
16			0.61E-2	5285
17			0.15E-3	3471

[a] Huzinaga basis set [16] for C,N,O:[51111/2111/1]; H:[311/1]
[b] the maximum of X,Y,Z Components is quoted
[c] CPU-times of a CONVEX C-210
[d] a DIIS extrapolation [11] has been performed after this iteration

ii.4)

For the iterative determination of the perturbation correction to the orbitals Pulays direct inversion in iterative subspace DIIS [11] can be applied successfully as shown in table [2] for a calculation of CH_3CHO.

DIIS reduces the number of iterations from 46 to 10 with the same level of accuracy for this example.

Table 2 - Convergence behaviour of the iterative solution of the IGLO equations with the use of DIIS [11] for a calculation of CH_3CHO (DZ+p[a]) basis N=65, M=16, K=7). The first order perturbation energy in u Hartree is plotted

iteration	$E_1(x)$	$E_1(y)$	$E_1(z)$
1	- 14.89425	- 15.52475	- 11.37018
2	- 16.67450	- 18.32755	- 11.81854
3	- 17.68408	- 20.24126	- 11.95525
4	- 18.28429	- 21.59950	- 12.00145
5	- 18.64746	- 22.58098	- 12.01808
6	- 18.86880	- 23.29742	- 12.02430 [b]
7	- 19.00413	- 23.82375	- 12.02823
8	- 19.08702 [b]	- 24.21208 [b]	- 12.02823
9	- 19.21862	- 25.33056	- 12.02823
10	- 19.21863	- 25.33056	- 12.02823

[a] Huzinaga basis [16]
[b] A DIIS extrapolation was performed after this iteration. DIIS is applied automatically if the Lagrangian multiplier (equation (6) in [11]) is smaller than a given threshold (here 1.0 E-12)

iii direct and parallel computation of the one electron
     operators and the $K_o^+(j)$ operators

When constructing the 1st order Fock operator, $K_o^+(j)$ operators are
needed only once in equation (5) to obtain their commutators with
the gauge functions.

Therefore it is possible to process the operators without using
intermediate storage on disk. Construction and processing of the
$K_o^+(j)$ operators is done by two different executables which can
communicate in the conventional sequential fashion via a normal
file on peripheral storage. For a multi-tasking computer it is
however possible to run the two programs in parallel comunicating
through a "pipe". Pipe and file are undistinguishable for the
FORTRAN program, and a UNIX operating system can run both
executables in parallel with the only provision that a file has
been changed to a pipe.

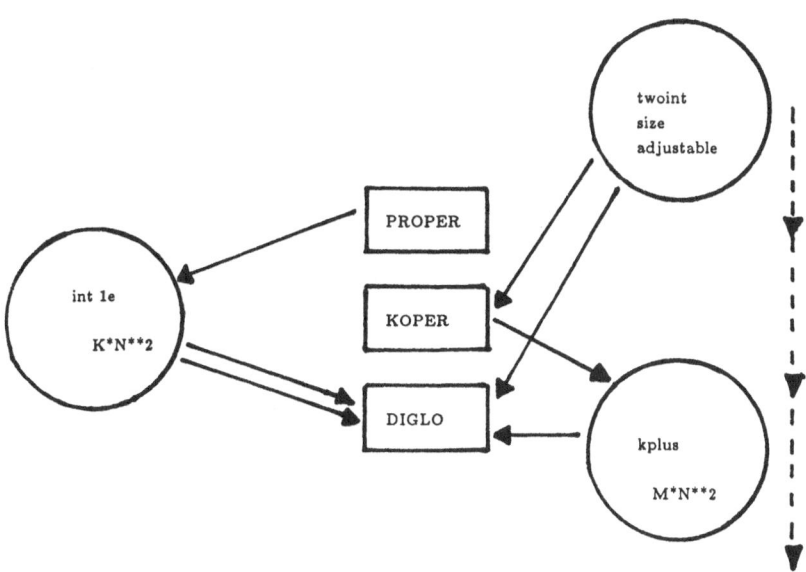

Fig.1 - Consecutive Processing

154

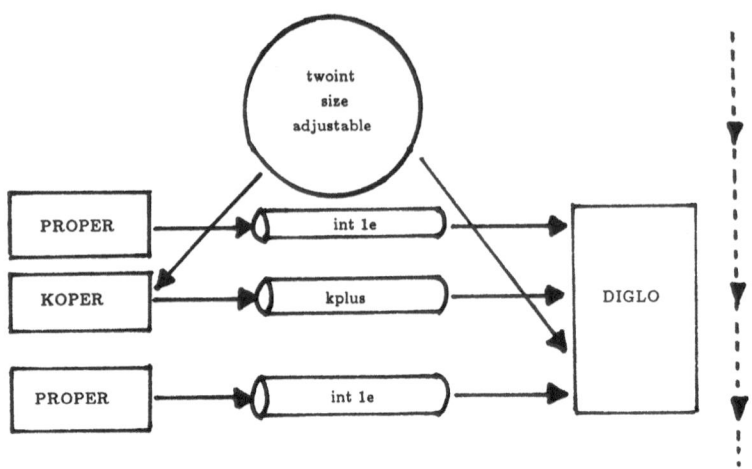

Fig.2 -Data Flow optimized Processing

Explanation for Figures 1 and 2

| | - executable program |

PROPER   - calculates the one-electron operators

KOPER   - calculates the $K_o(j)$ exchange operators

DIGLO   - does everything else, mainly solving the iterative
              IGLO equation and analyzing the perturbation
              correction.

              - is a file on disk

              - is a named pipe

int le   - contains the one-electron operators

kplus   - contains the $K_o^+(j)$ operators

twoint   - contains the two-electron integrals and is a size
              adjustable

⟶   - data flow

- - -▶   - time flow

The same procedure can be applied to the program which calculates the one-electron integrals. Figure [1] shows the program structure for a sequential run of the programs while figure [2] demonstrates a parallel data flow optimized run with pipelined I/O.

Only for a small sized example (Acetaldehyde with DZ-basis set N=65, M=16, K=7) the amount of disk storage needed by other methods are available. Since the size of the two electron integrale file is adjustable [5], and since there is no K*N**2 or M*N**2 dependance of disk storage for the direct IGLO program, the disk storage is no severe bottle neck for large scale calculations as it is for other versions of this program [12, 13] or the LORG program of Hansen and Bouman [14]. The storage requirements are listed in table [3].

Table 3 - Storage requirements for a calculation of CH_3CHO (DZ + p[a] basis, N=65, M=16, K=7)

DIGLO parallel [b]	12.3 MB
DIGLO sequential [b]	16.8 MB
IGLO [c]	42 MB
LORG [d]	103 MB

[a] Huzinaga basis [16]
[b] this work
[c] Schindler and Kutzelnigg [1, 2, 12, 13]
[d] Facelli, Grant, Bouman, Hansen [14]

Table 3b - Storage requirements for a calculation of N Methylacetanilid (basis ii, N=308, M=40, K=21)

	DIGLO parallel		DIGLO sequential
twoint	288 MB	to	2147 MB
int 1e	-		106 MB
kplus	-		15 MB
scratch	13 MB		13 MB

In conclusion, large scale IGLO calculations on supercomputers are possible by

- effective two electron integral processing as in the program TURBOMOL
- reduction of disk storage space by pipelined parallel and data flow optimized processing
- vectorization of the time determining step

Applications of this program to large molecules ($Sn\ R_3^+$, $R = CH_3$, ..., $C_7 H_{15}$) will be discussed in a forthcoming paper [15].

Aknowledgements

This work was started from a cooperation between the BAYER AG, Leverkusen, and the Lehrstuhl für Theoretische Chemie, Ruhr-Universität Bochum, and continued at the CONVEX Computer GmbH. The author thanks Priv.Doz.Dr. M.Schindler for initiating and stimulating this work and Prof.Dr. R.Ahlrichs for a copy of his program TURBOMOLE.

REFERENCES
----

[1]   U.Meier, C.v.Wüllen, M.Schindler, to be published

[2a]   W. Kutzelnigg, Israel J. Chem., 19, 193 (1980)

[2b]   M. Schindler, W. Kutzelnigg, J. Chem. Phys., 76, 1919 (1982)

[3]   W. Kutzelnigg, U. Fleischer, M. Schindler, in: NMR, Basic
       Principles and Progress, Springer-Verlag, 1989 in Press

[4]   W. Kutzelnigg, J. Molec. Struct. (THEOCHEM), 202, 11 (1989)

[5]   M. Häser, R. Ahlrichs, J. Comput. Chem., 10, 104 (1989)

[6]   R. Ahlrichs, M. Bär, M. Häser, M. Horn, C. Kömel,
       Chem. Phys. Lett., 162/3, 165 (1989)

[7]   S. Foster, S.F. Boys, Rev. Mod. Phys., 32, 296 (1960)

[8]   M. Dupins, J. Rys, H.F. King, J. Chem. Phys., 65, 111 (1976)

[9]   C.C.J. Roothaan, Rev. Mod. Phys., 23, 69 (1951)

[10]   R. Ahlrichs, Theoret. Chem. Acta, 33, 157 (1974)

[11]   P. Pulay, Chem. Phys. Lett., 73, 393 (1980)

[12]   M. Schindler, W. Kutzelnigg, J. Am. Chem. Soc., 105 ,
       1360 (1983)

[13]   M. Schindler, W. Kutzelnigg, Mol. Phys., 48, 781 (1981)

[14]   J. Facelli, D. Grant, T. Boumann, A. Hansen, J. Comput.
       Chem., 11, 32 (1990)

[15]   U. Meier, M. Schindler, G.Schüürmann, to be published

[16]   S. Huzinaga, Aproximate Atomic Wave Functions
       (University of Alberta, Edmonton, Alberta 1971)

Computer Aided Protein Design: Three Dimensional Model Building of the Saruplase Structure

W. Straßburger, W. Winter, G. J. Steffens, W. A. Günzler
and L. Flohé *
Grünenthal GmbH, Center of Research, W-5100 Aachen, FRG

Abstract

Modelling studies of the three-dimensional structures of the saruplase-domains are presented. The model of the N-terminal EGF-like domain highlights amino acids residues which might be involved in interactions with saruplase specific receptors. The distribution of charged residues on the surface of the kringle-model is different from other kringle-structures. The model structure of the catalytic serine-protease domain points to surface loops, which surround the active site and may participate in interactions with plasminogen. Starting from the structures of the isolated domains a model for the entire enzyme is constructed which is compatible with experimental results.

1. INTRODUCTION

Knowledge of the three-dimensional structure of the plasminogen activator saruplase (unglycosylated single-chain urokinasetype plasminogen activator, CAS 99149-95-8) should give insight into structure/function relationships of this enzyme. These could be the basis for the rational design of interesting variants with altered properties. Unfortunately, all attempts to obtain crystals for structure determination by X-ray crystallography have failed. We have therefore used theoretical methods to deduce a plausible model for the saruplase structure.

Examination of the primary structure of saruplase, as well as that of other fibrinolytic

---

* present address: GBF, Ges. für Biotechnologische Forschung, D 3300 Braunschweig

U. Harms (Ed.)
Supercomputer and Chemistry 2
© Springer-Verlag Berlin Heidelberg 1991

enzymes suggests the existence of different structural domains [1,2,3].

The N-terminal sequence is homologous to epidermal growth factors (EGF), and is followed by a kringle-domain, and then a serine protease domain. Starting from X-ray diffraction data and NMR-studies on such structural domains we developed tentative structures for the isolated domains using standard computer-aided modelling techniques. These structures were used for constructing a model-structure of the whole saruplase molecule.

2. METHODS

Model building was performed using the program system WHATIF [4]. This program provides tools for mutating amino acids side chains and handling deletions and insertions based on a relational proteinstructure database.

Energy refinement and molecular dynamic simulations of the modelled structures were carried out using the GROMOS program library [5]. The molecular dynamic simulations were performed using a TRACE 7/300. Figures were created using programs CHEMX (Chemical Design, Oxford, UK) and RIBBON (N. Jansen and W. Straßburger, unpublished)

3. RESULTS AND DISCUSSION

3.1 Growth-factor-like domain

Modelling of the EGF-like domain was based on the structure of human epidermal growth factor which has been determined by a combination of two-dimensional NMR-spectra and restrained molecular dynamic simulations [6]. Exchange of amino acid side chains was done according to the sequence alignment shown in Fig. 1.

All mutations could easily be accomodated into the parent structure. The shortening of the loop between Cys 6 and Cys 14 (EGF-numbering) did not result in major changes of the main chain conformation.

The structural model obtained after energy minimization and 50 picoseconds of molecular dynamic simulation is depicted in Fig. 2. Appella [7] has recently suggested which amino acids may be involved in receptor-binding to saruplase-specific receptors. Four of these amino acids (Lys 23, Asn 27, HIS 29 and Trp 30) are clustered on one side of the beta-structure (19-31) of the theoretical model and thus seem predestinated for receptor binding.

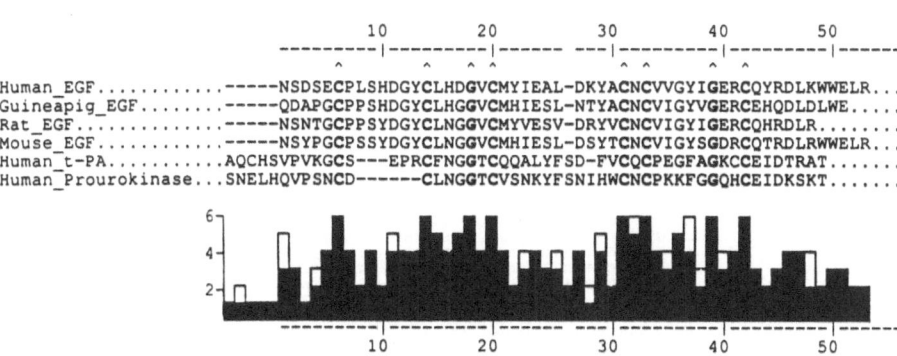

Fig. 1 Alignment of four sequences of EGFs isolated from different species and the corresponding parts of the tissue-type plasminogen activator (t-PA) and saruplase sequences. The diagram represents the number of identical amino acids (shown as filled columns) and of similar amino acids (shown as open columns). All sequences were taken from the EMBL, SWISSPROT or NBRF sequence databases.

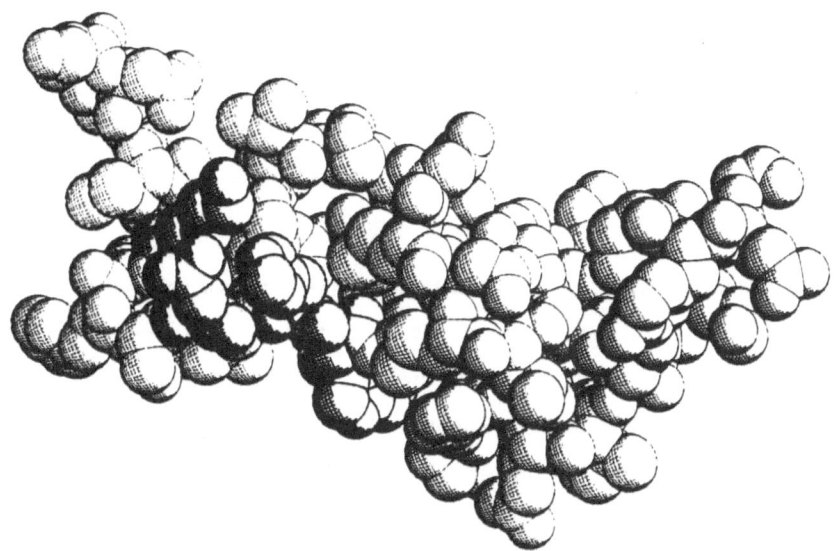

Fig. 2 Space-filling model of the EGF-like domain of saruplase. Amino acids possibly involved in receptor binding, are indicated by darker lines

3.2 Kringle domain

The tertiary structure of kringle fragments from different proteins has been determined in crystals and in solution [8,9]. Because no coordinates were available, we had to start by reconstructing the Cα-backbone of the prothrombin kringle-structure from the published stereo picture [8]. Using WHATIF, the side chains were automatically added and afterwards energy-refined, taking into account the published distances of the disulfide bridges.

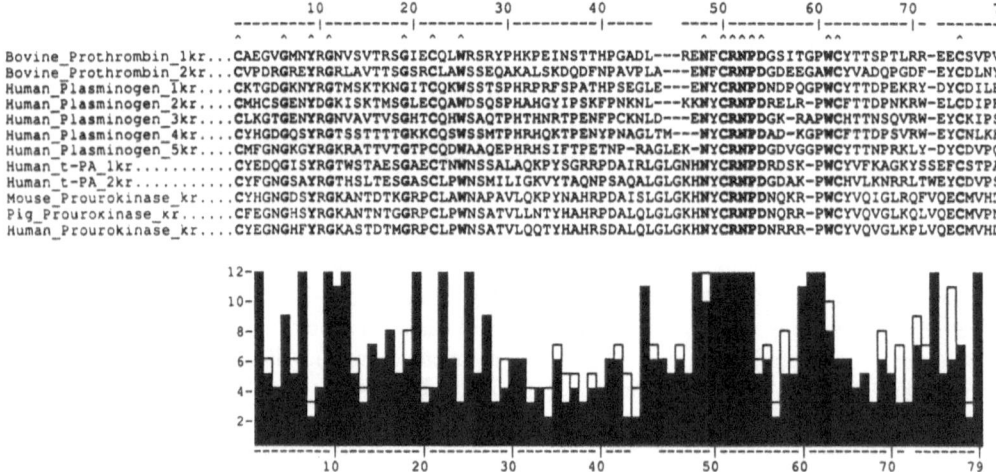

Fig. 3 Sequence alignment of kringle fragments from different enzymes. The diagram represents the number of identical amino acids (shown as filled columns) and of similar amino acids (shown as open columns).

Using the sequence alignment shown in Figure 3, the necessary amino acid replacements to the prothrombin sequence were performed to build model structures of the saruplase-kringle,and of kringle-4 of plasminogen (Fig. 4).

The general fold of the resulting structures is very similar but the distribution of charged amino acids on the surface around the conserved Trp 61 which may be involved in fibrin-binding [10] is very different. This may explain the differences in fibrin-binding propensity of these two kringles.

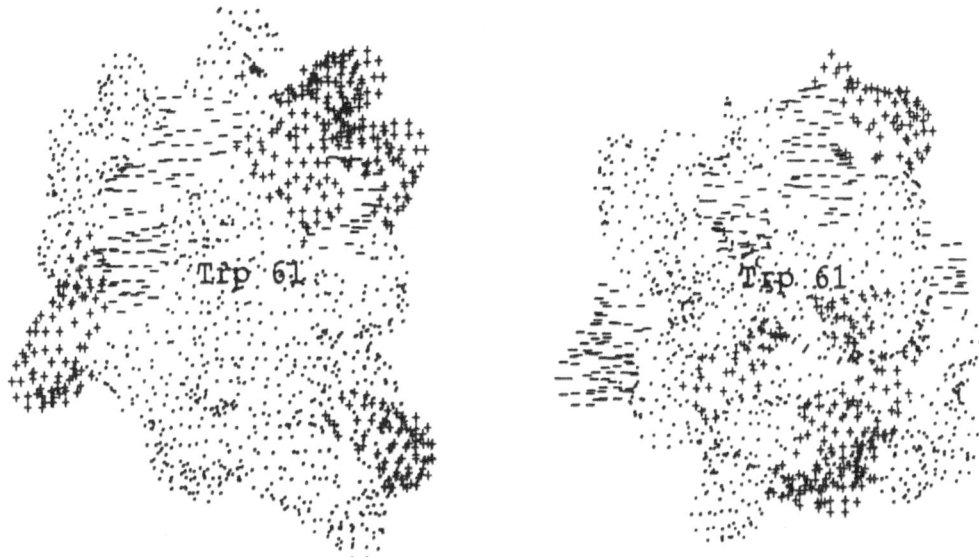

Fig. 4 Surface structures of kringle 4 from plasminogen (right) and the saruplase kringle (left). Plus signs indicate positively charged amino acids and minus signs negatively charged amino acids.

3.3 Serine-protease domain

Shortly after determining the primary structure of saruplase we developed a model-structure for the serine - protease domain, based on the known X-ray structure of other serine-proteases [11]. This model was further refined by molecular dynamic simulations using GROMOS, keeping the position of the Cα-atoms within the structurally-conserved regions constant.

As described in detail elsewhere [11] all insertions and deletions are outside the structurally-conserved regions and mostly located on the surface. The additional loops are positioned around the active site (Fig. 5) and thus possibly involved in interactions with plasminogen.

The loop-structures shown are energetically favoured but their real positions might be different and can only be deduced by experimental methods.

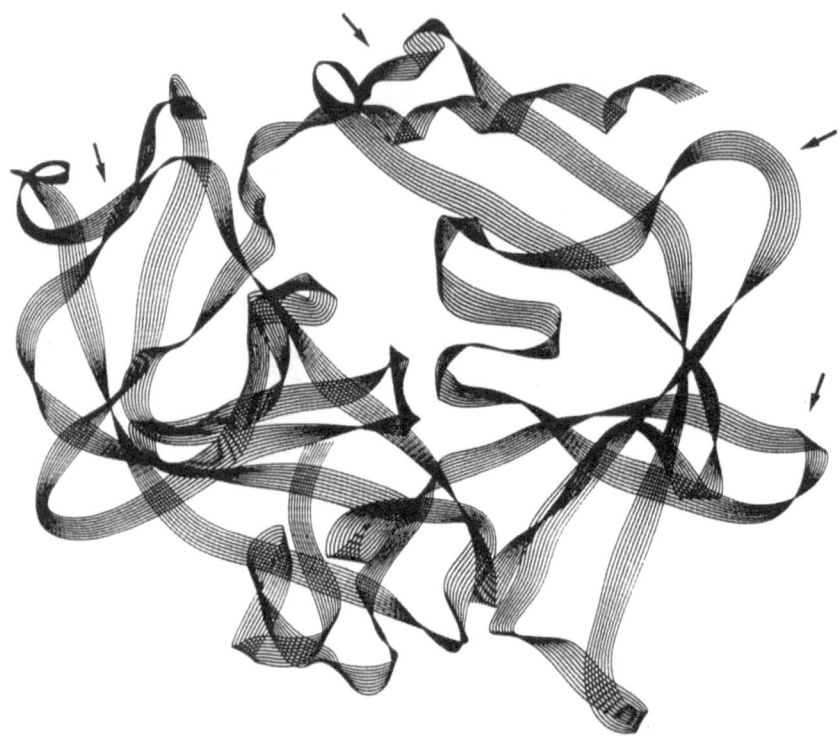

Fig. 5 Ribbon-diagram of the model of the serine-protease domain of saruplase. Arrows point to additional surface loops not present in the chymotrypsin structure.

3.4 Model structure of saruplase

Recent NMR-measurements of saruplase have shown [12], that all three domains move quite independently in solution. It seemed therefore justified to use the developed theoretical structures of the isolated domains to build a model for the entire enzyme. The arrangement choosen was suggested by low angle X-ray scattering measurments [13]. The conformation of the connecting peptides between the domains is arbitrary but compatible with the secondary structure perference of the constituent amino acids. The resulting structural model (Fig. 6) can only be regarded as a rough approximation to the structure of saruplase in solution. Further experimental results are certainly needed to refine this model.

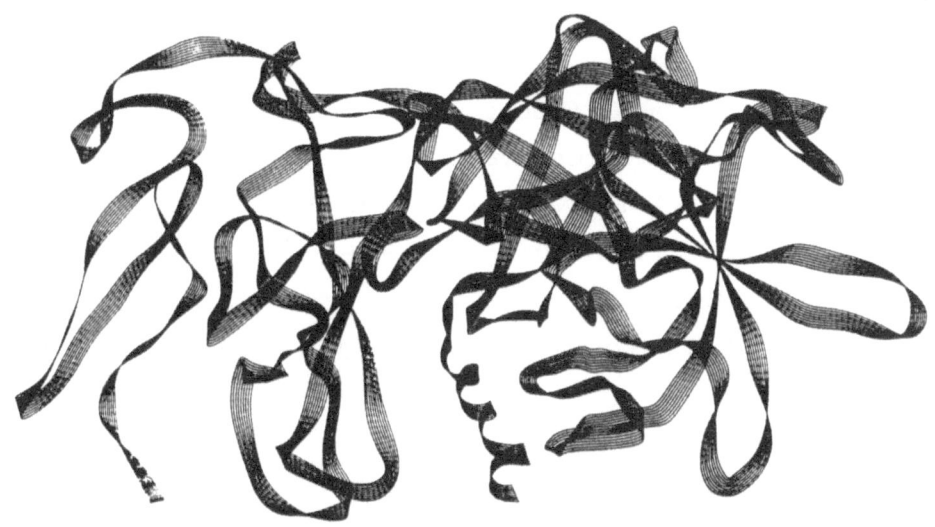

Fig. 6 Ribbon diagram of the model structure of saruplase.

Acknowledgements

The skillfull programming assistance of N. Jansen is greatfully acknowledged. E. Dahlke is thanked for assistance in the preparation of this manuscript. This work was supported financially by grant 03/8766A from the Bundesministerium für Forschung und Technologie (FRG).

4. REFERENCES

1. G.J. Steffens, W.A. Günzler, F. Ötting, E. Frankus and L. Flohé (1982)
   Hoppe-Seylers Z. Physiol. Chem. 363: 1043-1058

2. W.A. Günzler, G.J. Steffens, F. Ötting, G. Buse and L. Flohé (1982)
   Hoppe-Seylers Z. Physiol. Chem. 363: 133-141

3. W.E. Holmes, D. Pennica, M. Blaber, M.W. Rey, W.A. Günzler, G.J. Steffens and H.L. Heyneker (1985)
   Biotechnology 3: 923-929

4. G. Vriend (1990)

   J. Mol. Graphics 8: 52-56

5. W.F. van Gunsteren and H.J.C. Berendsen (1987)

   GROMOS-Library, Biomos, Groningen, The Netherlands

6. R.M. Cooke, A.J. Wilkinson, M. Baron, A. Pastore, M.J. Tappin, I.D. Campbell,

   H. Gregory and B. Sheard (1987)

   Nature 327: 339-341

7. E. Appella, E.A. Robinson, S.J. Ullrich, M.P. Stoppelli, A. Corti, G. Cassani and

   F. Blasi (1987)

   J. Biol. Chemistry 262: 4437-4440

8. A. Tulinsky, C.H. Park and E. Skrzypczak-Jankum (1988)

   J. Mol. Biol. 202: 885-901

9. B.C. Mabbutt and R.J.P. Williams (1988)

   Eur. J. Biochem. 170: 539-548

10. A. Tulinsky, C.H. Park, B. Mao and M.Llinas (1988)

    Proteins 3: 85-96

11. W. Straßburger, A. Wollmer, J.E. Pitts, I.D. Glover, I.J. Tickle, T.L. Blundell,

    G.J. Steffens, W.A. Günzler, F. Ötting and L. Flohé (1983)

    FEBS 157: 219-223

12. R.O. Oswald, M.J. Bogusky, M. Bamberger, R.A.G. Smith and C.M. Dobson (1989)

    Nature, 337: 579-582

13. S.A. Cederholm-Williams

    personal communication

V. I. Minkin, B. Ya. Simkin, R. M. Minyaev,
Rostov University, USSR

Quantum Chemistry of Organic Compounds

Mechanisms of Reactions

1990. XV, 270 pp. 66 figs. 35 tabs.
Hardcover ISBN 3-540-52530-0

This textbook on the application of ab initio and semi-empirical techniques for the analysis of organic reaction mechanisms is designed for chemistry undergraduates. The material is presented according to the mechanistic types: nucleophilic and electrophilic substitution, addition reactions, radical, pericyclic, proton and electron transfer reactions. Orbital and electrostatic models are used for structural correlations of interacting molecular systems along the reaction paths. Particular phenomenology with the basic theoretical principles needed to understand and predict chemical reactivity.

Springer-Verlag
Berlin
Heidelberg
New York
London
Paris
Tokyo
Hong Kong
Barcelona
Budapest

Z. B. Maksić, Zagreb, Yugoslavia (Ed.)

Theoretical Models of Chemical Bonding

Numerous experts have contributed to a four-volume-set of books presenting the broad spectrum of contemporary theoretical chemistry. The first volume reviews methods for representing molecules in theoretical models. The introduction of further properties, like electronegativity and relativistic effects, to individual molecules and solids is reviewed in the second volume. Theoretical models for the interaction of molecules with radiation is the subject of Vol. 3. The last volume bridges the gap to biological systems with the theoretical models for the properties of large molecules.

Part 1

Atomic Hypothesis and the Concept of Molecular Structure

1990. XXVIII, 324 pp. 40 figs. 51 tabs.
Hardcover ISBN 3-540-51578-X

Part 2

The Concept of the Chemical Bond

1990. X, 643 pp. 181 figs. 88 tabs.
Hardcover ISBN 3-540-51553-4

Part 3

Molecular Spectroscopy, Electronic Structure and Intramolecular Interactions

1991. X, 638 pp. 172 figs. 126 tabs.
Hardcover ISBN 3-540-52252-2

Part 4

Theoretical Treatment of Large Molecules and Their Interactions

1991. X, 458 pp. 104 figs. 52 tabs.
Hardcover ISBN 3-540-52253-0

> **4-Volume-Set**
> **ISBN 3-540-51741-3**
> Subscription price (valid only
> for subscribers to the
> complete work)

Springer-Verlag
Berlin
Heidelberg
New York
London
Paris
Tokyo
Hong Kong
Barcelona
Budapest